UNIQUE POWER SYSTEM PROBLEMS—SOLVED

Unique Power System Problems — Solved

O. C. Seevers, P.E.

 **THE FAIRMONT PRESS, INC.**
P. O. BOX 14227 • ATLANTA, GEORGIA 30324

Unique Power System Problems—Solved
by O. C. Seevers, P.E.

© 1982 by The Fairmont Press. All rights reserved. Printed in the United States of America. No part of this publication may be reproduced, stored in a retrieval system, or transmitted in any form or by any means, electronic, mechanical, photocopying, recording or otherwise, without the prior written permission of the publisher.

While every effort is made to provide dependable information, the publishers and author cannot be held responsible for any inaccuracy.

ISBN: 0915586-60-6

Second Printing: August 1983
Third Printing: June 1984

To my wife

Annie

Problems Solved

1 "Their telephones are ringing and won't stop." 2

2 "If it ain't grounded, it ain't dead." 4

3 What caused the right-hand 15-kva pot to burn up? 6

4 What is the value of 1 kw of peak system loss? 12

5 How do you get 360 volts single phase from a three-phase 240-volt bank? 14

6 "Their system isn't in phase with ours." 19

7 This may sound like a sloppy way to profile a line 22

8 He knew that our 69-kv voltage had run into his house and destroyed his 12-year-old TV set. 23

9 Determining line impedances the system analyzer won't tell you—doesn't it know? 30

10 Simplifying voltage loss and power factor calculations 32

11 Solving the I^2R riddle 36

12 What value to place on increased load as the result of decreased voltage drop 43

13 Understanding the interaction of currents, impedances, and power factors in a two-phase and neutral system 45

14 Little angles are to the engineer what blackheads are to the teenager 52

More Problems Solved

15 It was an old bank of three 25 Ckva capacitors
... Why did all three fail at the same time? 53

16 Rolling phases and mid-span clearances 55

17 Vector relationships 64

18 "What is the capacity of the 12-kv bus from which
you intend to serve our new plant?" 66

19 The three-phase motors were overheating 75

20 It's the connection directly to the coil in an
autotransformer that makes the difference 78

21 It's just a simple little old motor-starting problem 80

22 Determining the value of transformer losses in
rate analysis 88

23 Parable of the slop trough—or, improve your
grounds 92

24 "How much load will it take to melt that ice
off the 34.5 kv lines in an hour?" 99

25 "Mr. Neat"—the wires should have been messy 103

26 "Mr. Can't" was having trouble with badly
unbalanced voltages 105

27 Let's make sure we understand fully the
problem of rotation 113

28 Calculating the voltage dip on a two-phase 4.16-kv
line to supply a 50-hp, 240-volt, three-phase motor .. 144

UNIQUE POWER SYSTEM PROBLEMS—SOLVED

The trick was to gather information in bits and pieces from a multitude of sources, while concealing my abject ignorance. Most of the knowledge normally ascribed to a freshly graduated Electrical Engineer was concealed from me in a black cloud of confusion. It was a breathless race to find the answer before someone asked the question. It was just like kiting checks; I had to stay one step ahead of the authorities to keep from being revealed as a fraud.

I never did become an expert, full of confidence and ready answers. Instead, I've had to develop my skills in staying one step ahead of the questions. I've learned to say, "Let me check my notes. I believe I've got that one worked out." Then I go to work it out, make some notes, and triumphantly report my answer.

Another ploy is to profess that you know the answer to the problem but just don't quite understand exactly what the problem is. Keep asking for a more and more definitive statement of the problem. Sooner or later the answer will emerge and you smile knowingly, giving the impression that you knew the answer all along. You were just using this method to make sure the questioner fully understood the answer when it became apparent.

Always record and file away *both* the question and the answer. I've got some jim-dandy answers filed away for which I no longer remember the questions.

Thirty-seven years ago I started to work for a utility company as an iceman. Then I worked on a tree-trimming and right-of-way clearing crew. Then on a line crew. I've climbed lots of poles, surveyed lots of lines and prepared lots of work orders. This practical working experience has given me the confidence to win arguments with all classes of employees on technical questions, even when I was wrong.

It is customary for engineers of ripened years to yield to the urge to pass along hard-earned knowledge and experience to their juniors. This usually happens just before they begin to rot. And they profess motives of highest purity. Usually they claim that they feel compelled to pass along what their predecessors passed to them.

Not so in my case. What I've learned I've had to extract from textbooks and technical papers which were produced for a fee. All the free advice I've gotten was worth just what I paid for it.

My motive for writing this book is purely financial. If it's not worth a few bucks to you, it's surely not worth my time. Like all engineering economic determinations, the trade has to be beneficial to both of us or it can't be justified.

Let's take some examples of engineering problems I've tackled and see if you think it is worth your while.

1

The phone rang. It was for me. Lloyd, our district superintendent, was on the line, and he was mighty upset. "The people up at Frankville are up in arms. Their telephones are ringing and won't stop. I checked with the telephone company and they are upset, too. They say they have to use rubber gloves to work on their lines. Do you think there's any way we can be causing this?"

"Does it happen continuously, Lloyd, or just at certain times of the day?" I asked.

"It seems to die down after 5:00 p.m. until about 8:00 a.m.," he answered.

I thought a minute. Frankville was a rural settlement at the end of 20 miles of 12-kv grounded circuit. The only large customer was a rock quarry at the end of the line.

"Lloyd, send a man to the quarry and replace the fuse on the capacitor bank. It's blown."

He did, and it was.

The logic was this: The telephone company had bonded their cable to our grounded neutral wire some few years before. This was the only connection between our systems, except inductance. Since there had been no trouble until this occurrence and there had been no change in the character of our lines or loads, it was obvious that the problem was high voltage imposed on the telephone cable. Since the trouble coincided with the operating hours of the quarry, it was natural to suspect that their load, in conjunction with some change in the character of our lines, was causing the trouble. What could change to produce this voltage on our neutral? Something that caused a high neutral current. If the three-phase load current was unbalanced, the quarry would have to be the likely source. But they wouldn't stand for that condition for as long as the trouble had existed.

Then I recalled the capacitor bank at the quarry. If one fuse had blown, there would be 400 ckva, almost 60 amps of current galloping back over the neutral wire. That, times about 20 ohms of neutral impedance, resulted in about 1200 volts above earth ground, forcing current through all sorts of innocent equipment. No wonder the telephone linemen got out their rubber gloves!

Now, most of this 1200 volts was dissipated by effective grounding of our equipment, the customer's equipment, and the telephone company's equipment.

Why didn't our protective equipment deenergize our line? You guessed it. We used line reclosers with no provision for tripping on excessive neutral current. The oil circuit breaker at the substation had a ground relay, but it was necessarily set higher than the value of current returned to it by the relatively minor unbalance.

Why doesn't this sort of thing happen all the time? Well, you don't often have this much load, requiring this much ckva in capacitors, at the end of this much line.

My rapid reaction to this telephoned request for help was not due to my giant brain. It was due to the confidence I've developed over the years in my ability to analyze a problem calmly, using very basic tools of our trade. Nine out of ten complex problems are solved by application of Ohm's Law. Resist the urge to astound people by labeling the problem with some high-falutin' name. Simplify! Reduce the complicated circuitry in your mind to a simple series circuit with a voltage source pushing a current through a load impedance in series with the impedance of the source. Then you're on your way to a solution!

2

He was dead, all right! No doubt about that. His body was still almost too hot to touch three hours after the accident. Obviously, lots of current had passed through his body. 120-volt secondary voltage couldn't do that. This man got into 7200-volt primary. It had been raining hard and the man on the ground saw him clamber over the secondary on the pole and squeeze under the primary. Made it that time!

Three phases of 12 kv went straight through at the top of the pole. Two phases tapped off just below at 90°. They were fused on buckarms. The cutout door nearest the lineman was open, the fuse blown. At some point in the action, he had stuck his screwdriver into the bottom contacts of the cutout and called down to the man on the ground that the line was hot past the open cutout. The man on the ground remembered hearing the fuzzing sound.

The "dead" phase was the one he contacted. The question was: "How could it be hot if the cutout door was open?"

Well, there was an open wye-open delta bank connected to the two-phase line a few spans away. A three-phase motor was supplied from this bank. It was not properly protected for a single-phasing condition, so it remained connected to the line and continued to run, generating three-phase voltage. As a result, the "dead" secondary coil of the open wye bank remained energized and that, in turn, energized its primary coil. Although the tap line cutout was open, it was still "hot" on both sides, including the point our unfortunate lineman touched while standing on a grounded, wet surface. He short-circuited the primary coil and massive current was forced through his body.

This was a tragic way to reemphasize the old saw: "If it ain't grounded, it ain't dead."

It is interesting to inspect the current divisions that would result whether the motor coils were hooked delta or wye. They are depicted in the following figure.

UNIQUE POWER SYSTEM PROBLEMS—SOLVED

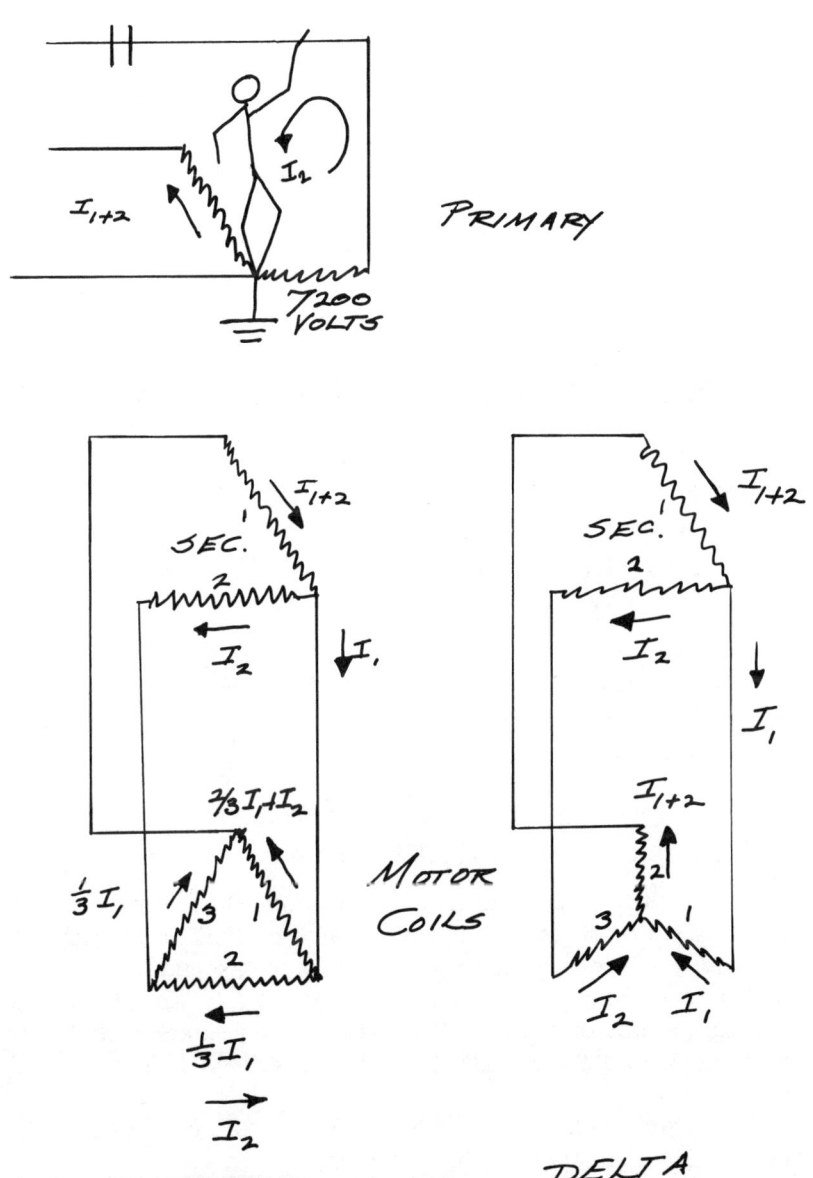

3

It was a three-transformer bank, hooked wye-delta. Two 15s and a 25. The 25 was center-tapped and grounded, and the lighting load was connected across it.

What caused the right-hand 15 kva pot to burn up? The superintendent was puzzled when he called. In his 37 years' experience he'd never seen one like this.

He assured me that lightning was not the culprit. He was disgusted to find that some ninny had grounded the high side "floating neutral." (Sometimes when a third transformer is added to an open bank, they forget to insulate the primary neutral.)

What threw him were the ammeter readings after they got the bank back in service. The 2.4 kv to 240 v readings were:

#1 — 25 kva   Primary — 5   Secondary — 50
#2 — 15 kva   Primary — 9   Secondary — 85
#3 — 15 kva   Primary — 11   Secondary — 100

The #3 spot was where the transformer failed. (See Fig. 1.)

The transformer secondary currents combined at the corners of the delta as follows:

1 — 2,   110
1 — 3,   90
2 — 3,   204

The secondary neutral read 25 amps.

Now, the first thing to keep in mind when someone calls and poses a riddle is that you must assume that all, or part, of the data you are given is screwed up. You can spend three weeks trying to solve a problem only to have the guy who asked the question admit sheepishly, "By the way, we rechecked that current figure and found out it should have been 185 instead of 85."

In this problem, you will notice that #2's 85 cannot combine with #3's 100 to produce 204 amperes.

Since the primary neutral is grounded, we know that the single-phase load will be supplied directly proportional to the respective transformer capacities and inversely proportional to their impedances. The impedances were:

UNIQUE POWER SYSTEM PROBLEMS—SOLVED

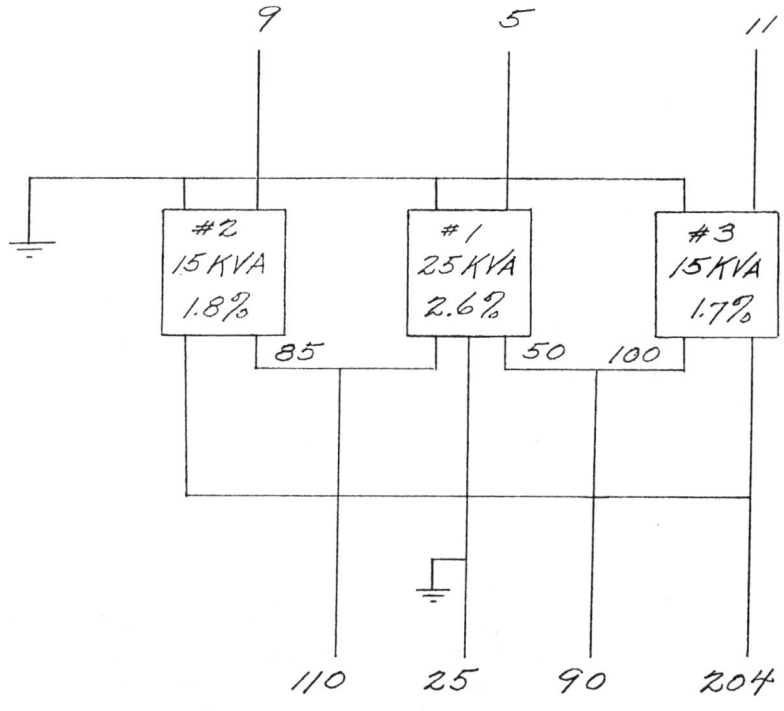

Fig. 1

#1 — 2.6%   #2 — 1.8%   #3 — 1.7%

Putting all three on a 25 kva base,

#1 — 2.6%   #2 — $\frac{25}{15}$ × 1.8 = 3.0%   #3 — $\frac{25}{15}$ × 1.7 = 2.84%

So, the single-phase load sees two parallel sources, one through #1 of 2.6% and the other through #2 and #3, in series, of 5.84% and should divide proportionately to 1/2.6, .384 and 1/5.84, .171, or 69% through #1 and 31% through #2 and #3 in series.

The three-phase load should divide equally.

But it didn't work that way, did it? What did happen?

I don't know for sure. But since things didn't follow the expected pattern, we have to look deeper to find something that could have caused this weird loading action.

I suspect that one of these units was on taps, applying 5% voltage to the closed delta, since the phase-voltages applied to the primary coils are held vectorially rigid by the grounded neutral, the unbalanced secondary voltage was impressed on the delta, resulting in circulating current.

Let's get some idea of how much this current would be. To do this, we divide the series delta impedance into the 5% voltage, or 12 volts.

The impedance of #1 can be found by noting that on short circuit the voltage drop is 100%. Therefore, the short-circuit current, assuming infinite capacity, would be 100%/2.6%, or 38.5 times full-load current, or 38.5 × 104 = 4000 amps. 240/4000 = .060 ohms.

$$\#2 \quad 100\%/1.8 \times 62.4 = 3470$$
$$240/3470 = .069$$
$$\#3 \quad 100\%/1.7 \times 62.4 = 3670$$
$$240/3670 = .065$$

Total delta impedance is .194.

12 volts/.194 ohms = 62 amps, lagging almost 90° from the voltage vector of the unit with the high voltage, the unit on taps. We immediately suspect that the circulating current is in opposition to the load current of unit #1. This explains how unit #1 can have less total current than either of the side units. It also indicates that I_c lags V_3 90°, so #3 is the unit on taps.

Assume 30° lagging load current. The vectors look like those shown in Fig. 2.

Let's check ourselves. We said that 69% of the single-phase load current should be fed by unit #1 and 31% by units 2 and 3, the current passing in series through both coils. Unit 2 has 35 amps and unit 3, 55 amps. Average, 45 amps. For reasons not germane to this problem, this unbalance of current through units 2 and 3 is typical of this transformer connection.

45 + 112 = 157 amps, or 100%
112/157 = 71%
45/157 = 29%. Close.

We took the easy way out on our diagram and solved for the resultant currents geometrically.

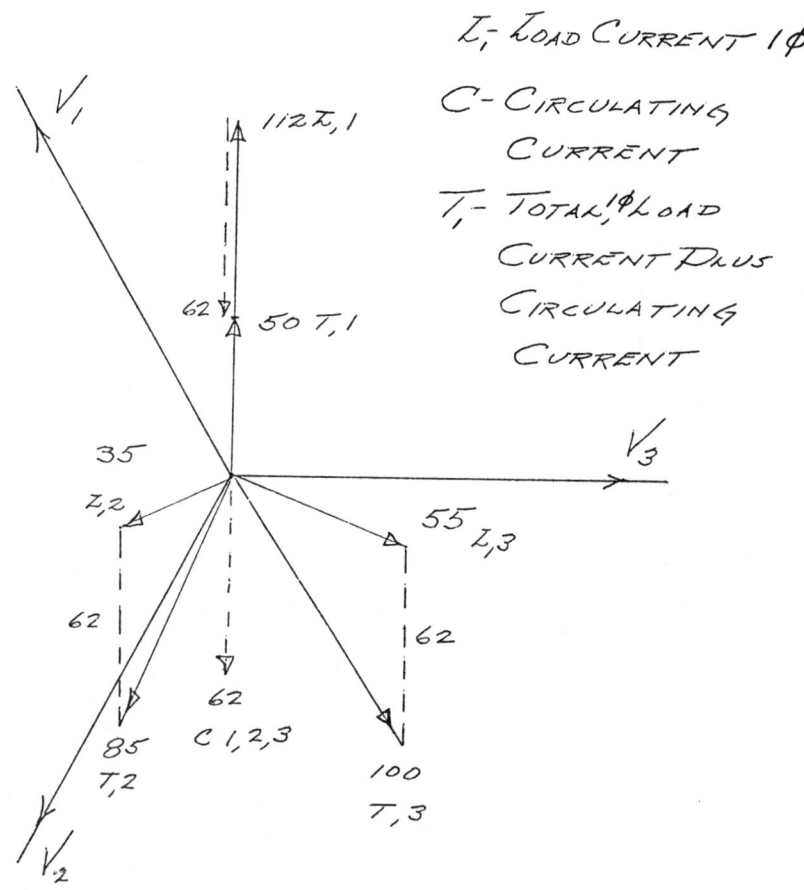

Fig. 2

Now let's get honest and practice our algebra, solving for $L_1 2$ and $L_1 3$.

($L_1 1$ is easy since it is the sum of $T_1 1 + C$, or $50 + 62 = 112$ amps.)

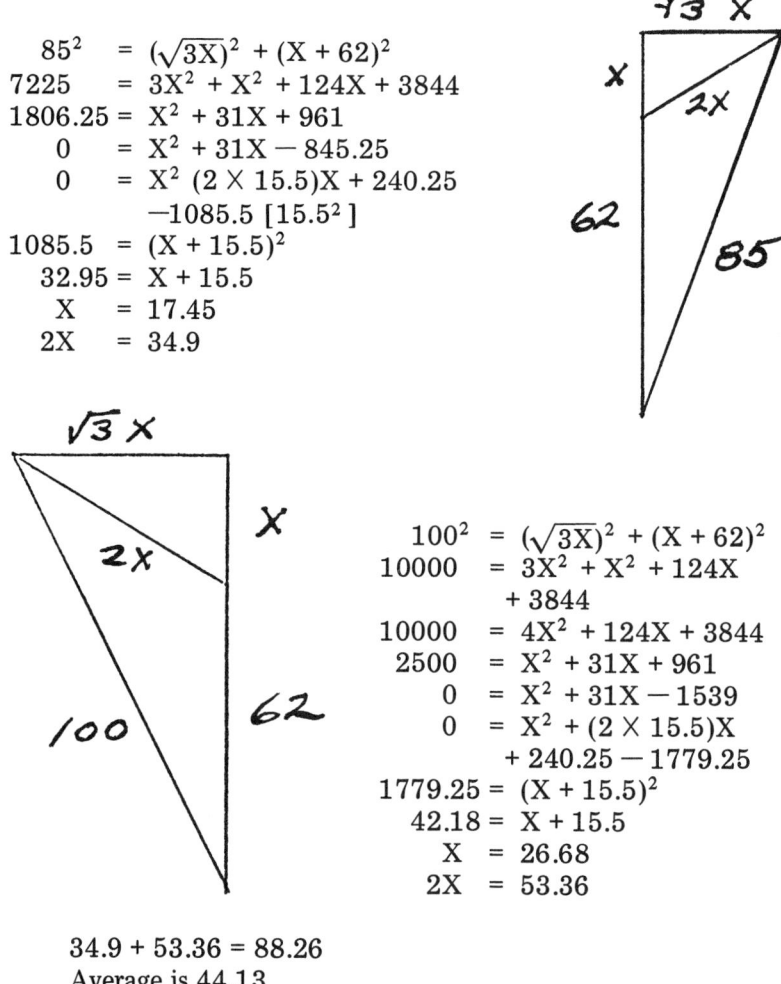

$85^2 = (\sqrt{3X})^2 + (X + 62)^2$
$7225 = 3X^2 + X^2 + 124X + 3844$
$1806.25 = X^2 + 31X + 961$
$0 = X^2 + 31X - 845.25$
$0 = X^2 (2 \times 15.5)X + 240.25$
$\quad\quad -1085.5\ [15.5^2]$
$1085.5 = (X + 15.5)^2$
$32.95 = X + 15.5$
$X = 17.45$
$2X = 34.9$

$100^2 = (\sqrt{3X})^2 + (X + 62)^2$
$10000 = 3X^2 + X^2 + 124X + 3844$
$10000 = 4X^2 + 124X + 3844$
$2500 = X^2 + 31X + 961$
$0 = X^2 + 31X - 1539$
$0 = X^2 + (2 \times 15.5)X + 240.25 - 1779.25$
$1779.25 = (X + 15.5)^2$
$42.18 = X + 15.5$
$X = 26.68$
$2X = 53.36$

34.9 + 53.36 = 88.26
Average is 44.13
44.13 + 112 = 156.13, or 100%
44.13/156.13 = 28.3%
112/156.13 = 71.7%

Close enough, considering we got our impedances and our current readings from a hungover lineman.

Now that my original suspicion has been confirmed, let's take a close look at the mechanics of circulating current in a transformer bank.

Remember, when you ground that "Y" point, the voltage vectors associated with the coils, both primary and secondary, become rigid. And when you connect three rigid-voltage vectors into a delta, it's not likely that delta will close. But the three coils are mechanically closed, so a fourth vector must be injected to offset the leftover voltage to reconcile the electrical diagram with the mechanical reality.

That fourth vector is formed from the voltage drop caused by the circulating current flowing through the three transformer impedances in series. This current is pushed along by the leftover voltage.

If the leftover voltage is very small, and it usually is, the circulating current is negligible. If it is sizable, as in our example, watch out! In this example, it was caused by someone stupidly hooking up a bank with one unit on taps. But the same thing could happen if the primary voltage balance was upset. This could happen if a regulator stuck, or a capacitor bank fuse blew.

None of these problems exists if you float the "Y" neutral point. The voltage vectors associated with the coils are free to swing and sway in any way necessary to keep the system balanced.

There is one more danger to expect if you ground the "Y" point neutral. Visualize a Y–Delta bank located a mile from the sub on a line which continues three miles on into the country. A sectionalizing fuse located between the sub and the bank blows. Call it "A" phase. The other two phase-to-ground voltages continue as before, retaining their secondary voltage relationships keeping two sides of the delta intact.

The third coil is dead as a source, but is kept energized by the other two in the same vector relationship as before. This voltage is transformed to primary phase "A," so normal voltage continues to appear across all loads connected to phase "A" past the blown fuse. The bank then sees this as a single-phase load and attempts to pump load current into phase "A." This load current circulates in the delta and is fed by the other two primary phases. This continues until one unit burns up or a transformer fuse blows.

If you've followed me closely to this point you've probably forgotten the problem we started with.

Remember, the right-hand 15 kva transformer burned up. We've validated the 100-amp secondary coil current reading given by the lineman. He probably didn't catch it on maximum peak, but 25 kva load is enough to cook a 15 kva pot if it hangs on long enough.

After the #3 unit (which was on 5% tap) was replaced and the primary neutral refloated, the currents all returned to normal. But the superintendent never did. Do they ever?

4

The value you place on 1 kw of peak system loss is the most important single figure you will ever calculate. On the accuracy of this one figure hangs thousands of economic decisions. Can you afford to replace a #2 ACSR circuit with 266.8 Mcm? 397.5 Mcm? Can you afford *not* to? That's more important. Is it economically supportable to convert a 4 kv system to 12 kv? 22 kv? Can a transmission line be operated one more year at 69 kv before converting to 138 kv? What percentage overload can you impose on a substation transformer before it becomes economic foolishness?

It has always seemed to me that the value of one kw loss, on peak, was perfectly obvious. Yet I've seen many intricately involved articles supplying totally unfathomable charts and equations. The one thing you had to know before you could use all this information was the value of one kw of peak loss. Big deal!

Now let's use some common sense and get an answer to this thing.

Every peak kw of loss on your distribution system has to be generated, transmitted and distributed. How can a generator, transmission line or distribution line be expected to tell whether the last kw loaded on its back is going to a paying load, or is going to be dissipated as heat and lost into the atmosphere? After all, that's load, just as much as any metered kw. Only you don't get paid for it. But you do have to provide just as much transmission capacity for it as for a metered kw; just as much distribution capacity; just as much generating capacity.

You can easily determine from your annual reports how

much your system peak demand was increased, say, in the last five years. Well, how much did your company invest in generation during that time? How much in transmission? Distribution? From this you can determine the investment corresponding to a peak kw. The annual cost of this investment is simply this investment times your company's fixed cost rate. (Inflation? Factor it in as you see fit, but my basic method is valid, despite the politicians.)

One other cost associated with one kw peak loss is the fuel cost to provide the kwh's which will be consumed by the one kw peak loss during the year.

Let's take an example: Assume your peak load increased 1,000,000 kw in the past five years. During that time you spent $120,000,000 on generation; $70,000,000 on transmission; and $110,000,000 on distribution. Assume 20% annual fixed costs.

Generation	$120/kw × 20% =	$24/yr/kw
Transmission	$ 70/kw × 20% =	14/yr/kw
Distribution	$110/kw × 20% =	22/yr/kw
Total annual cost	$ 60/yr/kw	

One kw loss at peak load consumes 1 kwh/hr. If the peak lasted all year, 1 kw loss would consume 8760 kwh's. Assume 60% load factor on your system. This will result in about 45% loss factor; 8760 kwh × 45% = 3940 kwh/kw/yr. Assume $.004/kwh fuel cost; 3940 × $.004 = $15.76/kw/yr.

Now you see that the value to you of one kw peak loss on your distribution system is about $76/yr. If you are working on a transmission problem, it is worth only $54/yr.

It is amusing, and disgusting, to read that generation costs should not be included because the decision to add a 500 Mw generator is made without any consideration for the relatively small portion of the load that is made up of losses. This is not too bright a statement, because the reduction of just one kw peak loss *postpones* the construction of one 500 Mw generator a definite amount of time, no matter how small that time figures to be. And that postpones the investment just that long, and saves the fixed charges on the entire investment for that small amount of time.

In our example, the load increased 200 Mw/yr. That's one added 500 Mw unit every 2.5 years; $120/kw × 500,000 kw = $60,000,000.

14   UNIQUE POWER SYSTEM PROBLEMS—SOLVED

If we could reduce losses by 100 Mw the 500 Mw addition could be postponed until 500/400 × 2.5 yrs = 3.125 yrs. That's easy to see.

It's not so easy for most folks to see that if we reduce losses by only one li'l ol' kw that the addition can be postponed until 500,000/499,999 × 2.5 yrs = 2.500005 yrs.

That postpones a $60,000,000 investment by .000005/2.5 × $60,000,000 = $120, of course. And since this 1 kw of reduced loss will never appear on the system, all future generator additions will be similarly postponed. This 1 kw will never require generator capacity, transmission capacity, or distribution capacity, nor will it consume any fuel.

The annual savings on this averted $120 investment will be as previously calculated.

By the time you read this, inflation will have made my assumed cost look downright silly. But the logic will still be good. The value of reducing losses by 1 kw will be lots more, but we won't be able to afford the rent on the money we'd like to borrow to do the work required to reduce the loss. Catch 22.

5

Time to get back to the "old man" tales. I'm really not so old. I do forget to zip up from time to time, but I haven't got to the point where I forget to zip down.

Here's a worthwhile exercise you will enjoy posing to the smart guy who is always showing you up: How do you get 360 volts single phase from a three-phase 240-volt bank?

First, let's go with a Delta-Delta. Use four transformers with two of the primaries in parallel. O.K.? Now, hook two secondary coils as usual, but reverse the third coil and hook the fourth in series with the third. Like this:

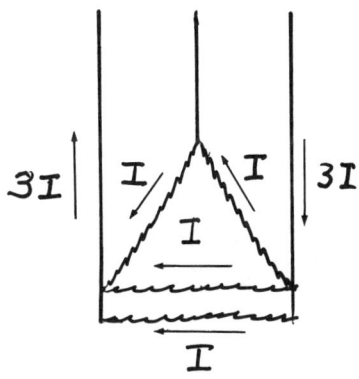

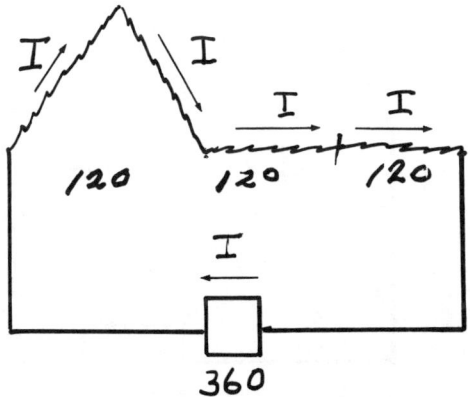

(3I) V Primary = (3V) I Secondary

Now, let's go to "Y"-Delta.

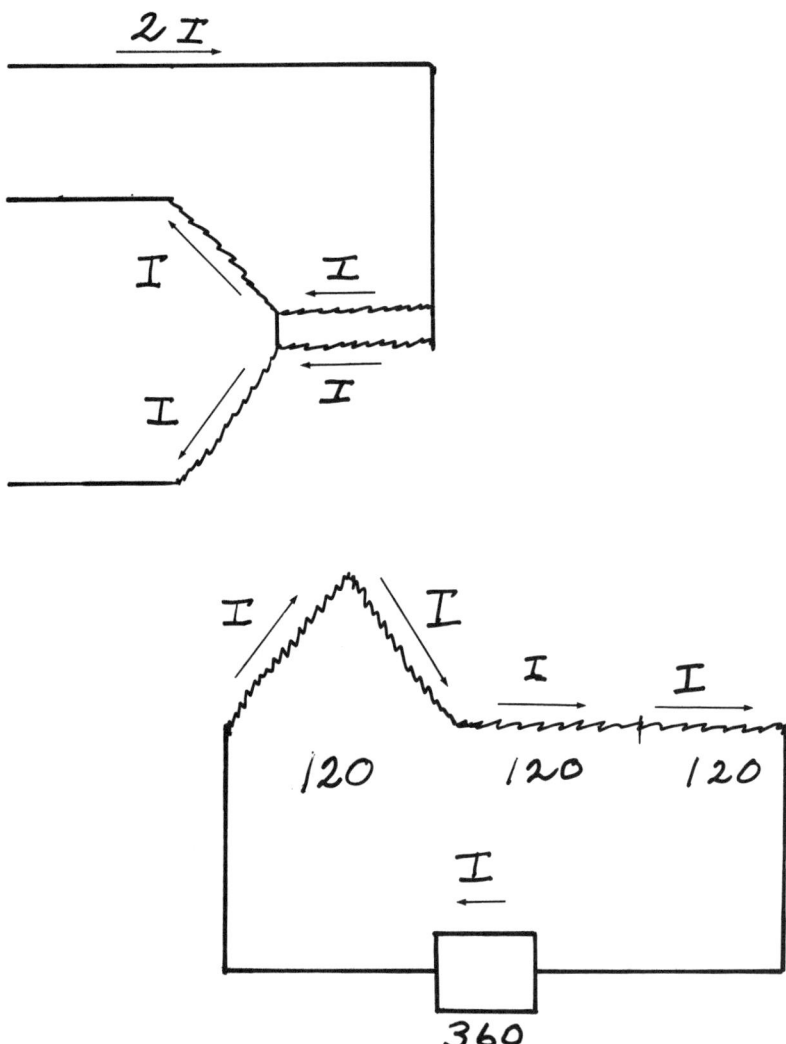

(2I)V + (2I) (.5V) = (3I)V Primary = (3V)I Secondary

No problem so far. Now let's get 240 volts single phase from the same bank hook-up. But this time let's use only three transformers.

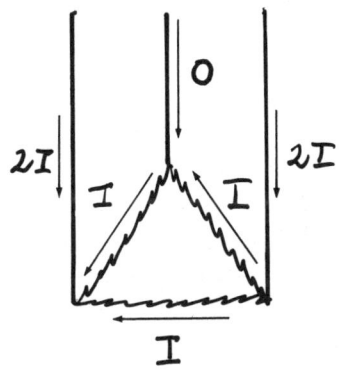

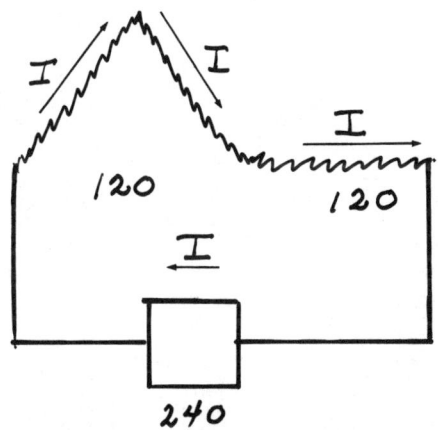

(2I) V Primary = (2V) I Secondary

Now the plot thickens, and hardens, and sets—and sets. Hook the primary in the above diagram in wye.

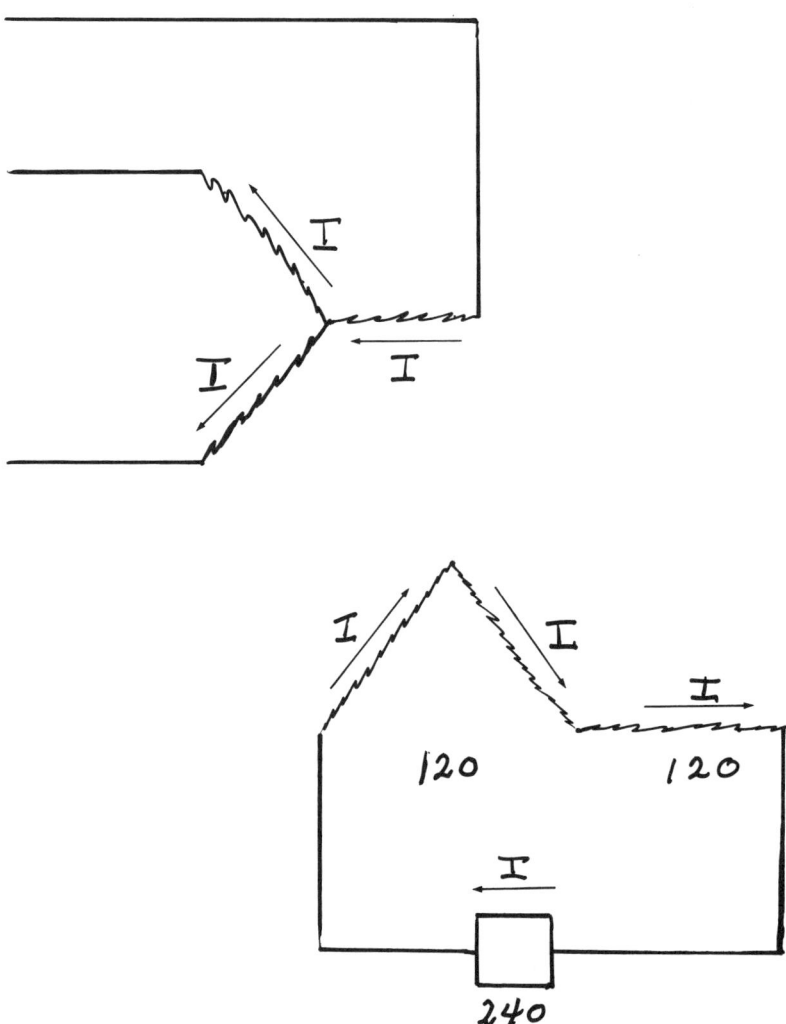

Note that the single-phase current in the "Y" enters the neutral point with a value of unity and comes out doubled.

Here's where you stop, stand back and smile knowingly, daring your friend to figure this one out.

Actually, only a tiny current will flow because the extra "I" must be supplied through the leakage reactances of the two transformers to the left. You have to return to the equivalent diagram of a transformer. The leakage reactance is so high that only a trickle of current will flow.

From this you learn that you can't use simple transformer diagrams without keeping in mind that they are accurate only when balanced, ordinary conditions prevail. They can lead you down the garden path on a problem such as this one. Shame on you!

6

If you want to get there in a hurry, hop on an electron. Electricity is about the fastest thing there is. Don't ever try to beat it to the draw.

Nevertheless, at 186,000 miles per second electricity is moving at a finite speed. In fact, 186,000 miles per second is so slow we can actually observe the effects of this finiteness.

Once upon a time, I got a call from our substation superintendent who was at a substation about 100 miles away. He was extremely confused. He was in the process of tying our 69 kv system together with the 69 kv system of another utility.

"Their system isn't in phase with ours," he said, "and I can't get them to go together."

"Did you phase out before you tried to close the switch?"

"Yes, that's what's so peculiar. We read almost 10 kv across each pole of the switch and we're afraid to close it. I could understand a 30° phase shift, but this is not like anything I've ever seen."

"Hold on a minute and let me think." I did some quick figuring and then asked, "Aren't we tied together with them already over another route?"

"Yes."

"Their generator is on that end and ours is on this end of 100 miles of line?"

"Yep."

"And if you close that switch down there it will complete a circular path?"

"That's right."

I made a couple more calculations and then I got back on the phone. "It's O.K. Go ahead and close that switch, but don't be timid in closing it."

He did just that and the two systems went together just fine. No trouble at all.

You've probably already guessed the answer from the clue I gave you at the start. Yes, it takes electricity some time to go 100 miles. While the two generators were locked in together by means of the existing tie, the voltage still had to travel 100 miles over our 69 kv line to our side of the open switch.

On the other side of the switch the voltage was "already there" by virtue of the other system tie.

Now visualize this: If alternating voltage were one cycle per second, this one cycle would stretch out over 186,000 miles. Since our voltage is actually 60 cycles per second, 60 cycles stretch out over a distance of 186,000 miles. And one cycle stretches out over 3100 miles.

There are 360 degrees in each cycle, so one degree stretches out over a distance of 8.6 miles. We were dealing with a distance of 100 miles, so the voltages on the two sides of that switch were out of phase 100/8.6, or 11.6 degrees. See diagram on facing page.

The voltage appearing across the open switch poles was the difference, 40 kv times the sine of 11.6°, or 8.04 kv. This was enough to make that switch spit real good as it was closed, but the voltage difference disappeared just as soon as the contacts came together. Electrically, the 100 miles just evaporated.

But I wouldn't want to be standing under that switch if someone tried to reopen it. You can bet it would beller like a mad bull before the ensuing arc would break!

UNIQUE POWER SYSTEM PROBLEMS—SOLVED

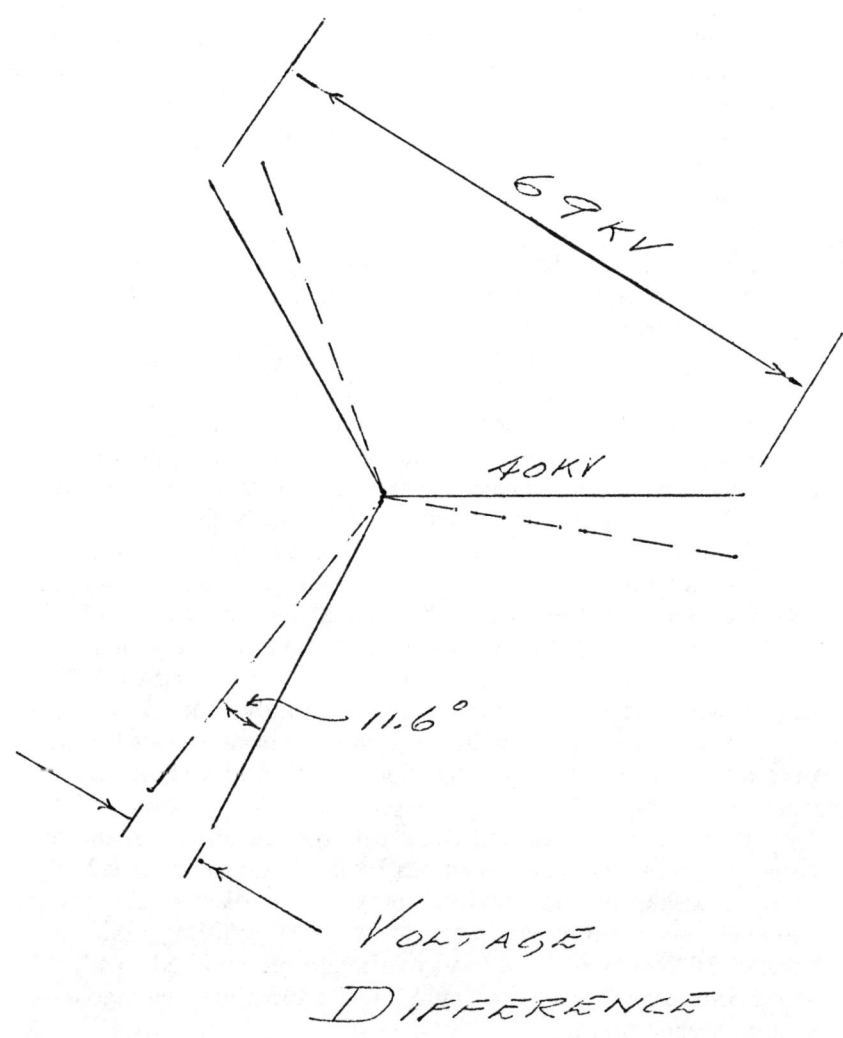

7

There were only two of us and we had to survey miles and miles of line with no help at all. So we had to devise some unique surveying methods to get the job done in a hurry because my buddy had a woman he was courting in the afternoons while her husband was at work. So we had to complete a day's work in the morning with as little effort as possible on his part to keep him in shape for the strenuous endeavors of his romantic p.m.'s.

Somebody had to walk the entire line, of course. My buddy was the boss, so I was elected. Besides, I could rest up all afternoon working up our notes.

The lines had to be profiled. Romeo ran the transit, which was equipped with a stadia circle, and I carried a 16-foot stadia rod and a surveying notebook. He had a notebook, too.

He set up the transit and picked out a spot on line, maybe a mile or two away. Sometimes we drove to that spot first and hung a big white rag on it so that he could easily site on it.

Then I walked to the first break in the ground contour and he read stadia and vertical angle. He assumed a transit height of five feet and set the middle horizontal hair on the five-foot mark on my rod. Actually, we were profiling a line which was suspended five feet above the ground. Then he jotted these figures down beside point number 1. Then I moved on to point 2, etc., until I got so far away he could no longer make out the figures on my stadia rod. He signaled to me and I took over as notebook-keeper.

From here on I turned the stadia rod upside down and he took all his vertical angle readings by lining up the middle horizontal crosshair on the top (bottom) of my stadia rod. He noted the angle in his book. At this point we were profiling a line suspended 16 feet above the level of the ground and had to adjust our calculations by subtracting 11 feet from the calculated elevation of each point.

The distances we read this way: Romeo set the top horizontal crosshair on the top (bottom) of the rod and I slid a red handkerchief up and down the rod as he signaled me to raise it or lower it, until he got my handkerchief set even with his bot-

UNIQUE POWER SYSTEM PROBLEMS—SOLVED 23

tom crosshair. Then I read the figures off the rod and jotted it down beside the point number in my book.

When I got over 1600 feet away, he began to read half stadia. Sometimes, he couldn't see all the way to the bottom of my rod even before I got 1600 feet away and I had to mount my red handkerchief on a stick to reach up to the point where he was reading half stadia.

We could often profile over a mile of line at one setup in this manner.

This may sound like a sloppy way to profile a line, but we built lots of lines surveyed in that unorthodox manner and they're still standing. We never goofed once. Couldn't afford to with so much riding on finishing Romeo's day's work by noon every day.

It was tough on him standing in one spot all day in the bright sun.

But I noticed he had always cooled down by the next morning.

8

It's not true. It just can't be true. I refuse to believe that every time a regulator sticks on raise, or we have an unusual fault condition, that the local manager goes door-to-door soliciting claims for damaged appliances!

Nevertheless, his son-in-law and his neighbors all seem to come out of our misfortunes with new TV sets. I've even had claims from customers whose service came from a 12 kv system when our trouble was on the 4 kv system.

Many times people file claims in all honesty because some appliance failed at the same time our system acted up.

My next story is about just such a case.

We were stringing 69 kv conductors in on 60- and 70-foot poles. Underbuilt on these poles was an energized 4 kv circuit which fed all the customers in that neighborhood.

When quitting time came, the ends of the three 69 kv conductors were bundled together and brought to a point about ten feet above ground on one of the poles. The three wire grips were attached to a chain which encircled the pole. Now, this chain was in contact with the pole ground, so the 69 kv conductors were inadvertently grounded. No one realized this at the time. This meant that the grounded 69 kv conductors passed from this point up between the phases of the 4 kv circuit to the top of the next pole.

Naturally, a freak storm broke out during the night with 50-mile-an-hour winds. It blew the 69 kv conductors over into contact with a 4 kv phase.

This caused several oil circuit-breaker operations on the 4 kv circuit and momentary outages to the customers in the area.

Hundreds of people were affected by this mishap, but only one had ever worked for an electric utility, so he knew all about electricity and quickly grasped that our 69 kv voltage had run into his house and destroyed his 12-year-old TV set. He saw the lights flare up brightly each time the breaker reclosed, sending 69,000 volts coursing through his rather wonderful light bulbs and his not-so-wonderful TV set. The bulbs survived. The TV set died.

When you get a ridiculous damage claim, such as this one, you smile to yourself and think private thoughts, but you never betray these impressions to the claimant. He's probably honest in his conviction that you caused his trouble. Good public relations demands that you hear him out with consideration and compassion without admitting a thing.

Good common sense dictates that you listen real good to get all the facts, because, although his claim looks ridiculous, he may just be right!

In this case, we patrolled the 69 kv conductors to be sure that the only energizing came from the obvious 4 kv contact. That checked out. Only 4 kv, 2.4 kv to ground.

But the man's set did die, and his light bulbs did flare up. True, his equipment was subjected to some overvoltage from some source.

There was nothing to indicate that the 2.4 kv voltage had increased. It was suggested that perhaps during the fault, the neutral was displaced. So we went to work on that theory.

If our friend was served from the 2.4 kv phase which had come in contact with the ground, it is obvious that his voltage would have tended to collapse, not increase.

But if he was served from one of the other phases (as he was), it is possible the fault current which coursed through the neutral and the ground on its flight back to the substation ground (this current, times the ground resistance) might possibly produce enough voltage drop to displace the neutral in the vicinity of the fault. This could impress overvoltage on transformers served from the two unfaulted phases. The big question then becomes, "How much is it possible to displace the neutral in such a situation?"

First, let's take a look at the circuit (Figure 1 on page 26).

To help you visualize our problem, note that if the sum of the impedance of transformer "A" and phase wire "A" is equal to the impedance of the return path (neutral wire in parallel with earth ground return), the fault current will produce a voltage drop from the sub ground out to the fault equal to the voltage drop it produces from the fault back to the sub ground. And if this were the case, which we hope it ain't, the neutral in the vicinity of the fault would be displaced half of 2400 volts, or 1200 volts. It would actually crawl out the faulted phase half way. See Figure 2, page 27.

Phase B and C would then be subjected to about 3170 volts. A 20/1 transformer would then put out about 158 volts.

We should not overlook the possibility that the lamp receptacle, and other receptacles, the cases of ranges and washers, etc., may be grounded to pipes which may be undisturbed by our neutral shift. Or they may not be shifted from zero potential nearly as much as our neutral. If so, a maximum of 1200 volts can appear between the appliance and the receptacle or washer frame. This could result in a sparkover and damage the equipment. Of course, we do all we can to tie all these grounds together to avoid just this. If our neutral falls away from zero potential we want all other associated grounds to fall away an equal amount so that no difference of potential will appear.

Did you ever hear your wall receptacles spit when lightning "ran into" your house? Lightning is sometimes just too much for our grounding system and enough potential difference appears to cause the sparkover situation I've just described.

Back to our original problem. Let's investigate the impedances of our system to see just what neutral shift we can expect.

First, the substation transformer impedance is usually

UNIQUE POWER SYSTEM PROBLEMS—SOLVED

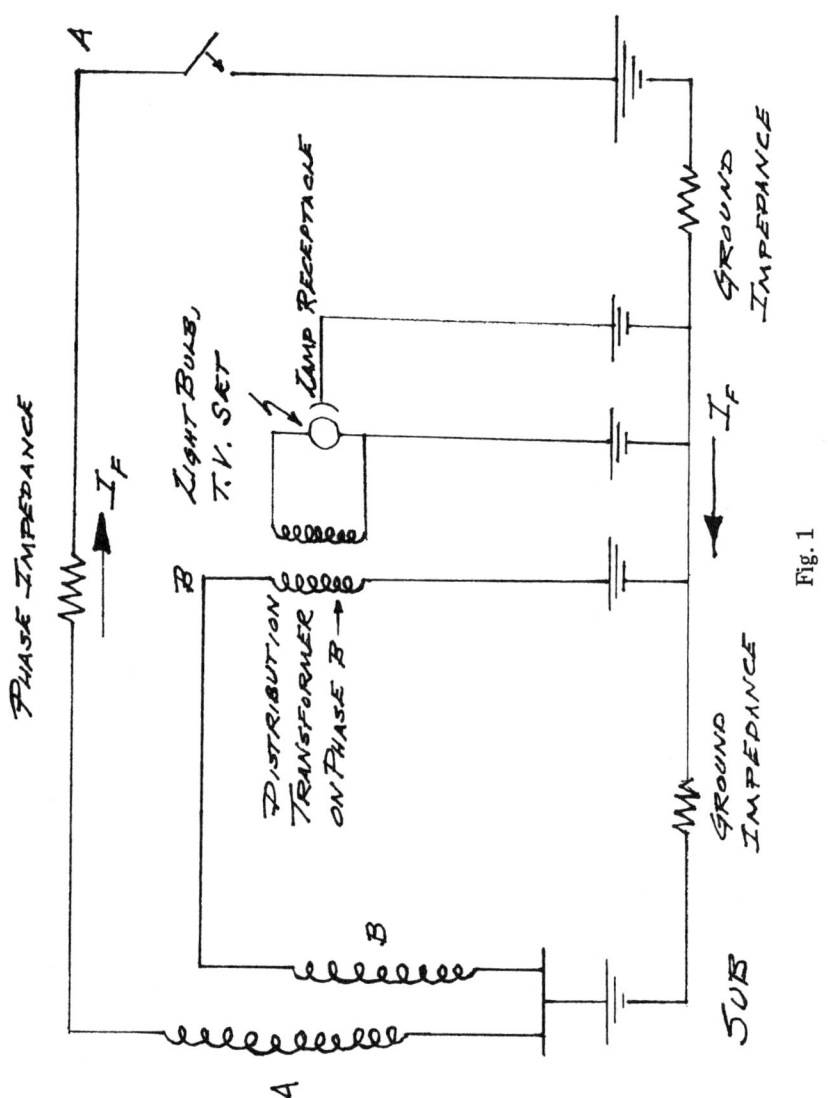

Fig. 1

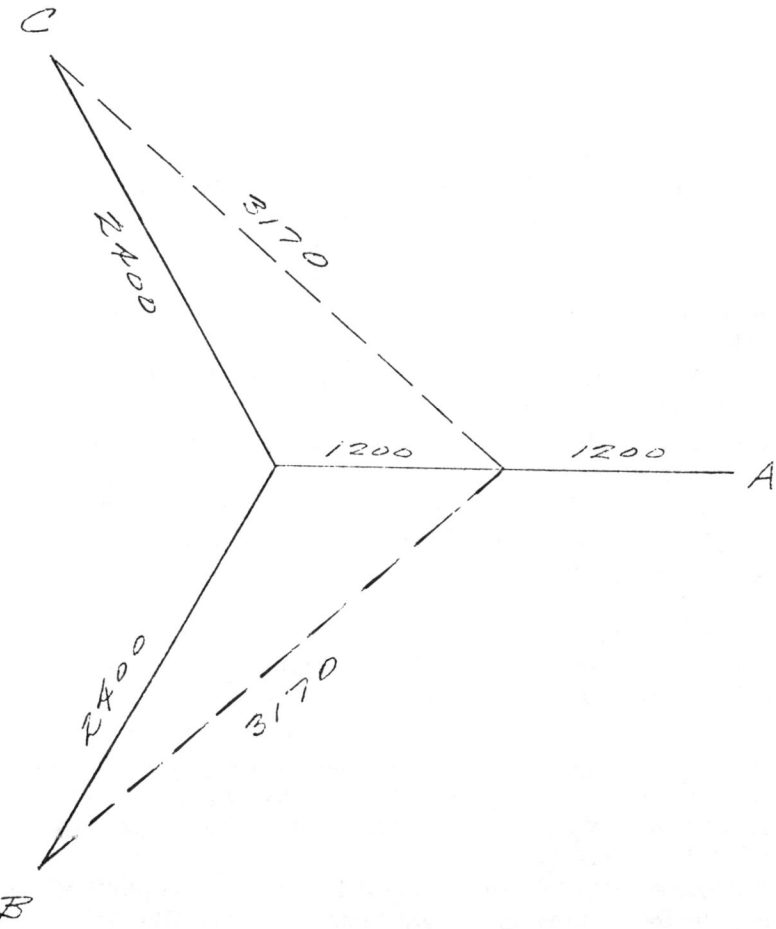

Fig. 2

about 7%. This means that if we short out the 2.4 kv coil and increase the primary coil voltage until the fault current is equal to full load current, we will find that it takes 7% voltage, or 175 volts on a 2.4 kv base. By dividing 175 volts by full load current, 1000 amps in the case of a 7500 kva unit such as we had here, we see that the impedance of one phase of the transformer is about .175 ohms, which is negligible in this problem.

The fault was about two miles from the sub, so the phase impedance was about two ohms.

The ground return measured about one ohm. Now, bear in mind that with an interconnected neutral system, multigrounded, it is highly unlikely that the ground return path could ever be expected to amount to more than half of the phase impedance, and we see that that is just about what we had here. So the neutral could not shift over one-third of the way over. That gives us only 800 volts neutral shift (Figure 3, facing page).

Phase B and Phase C will then be subjected to 2890 volts. A 20/1 transformer would then put out a maximum of 145 volts.

A momentary input of 145 volts is just not enough to cause damage to healthy equipment. This is borne out by the fact that hundreds of other customers in the area of our fault reported not one case of equipment damage.

However, 145 volts may be just what a 12-year-old TV picture tube was waiting for to bid a fond farewell to the Ponderosa and Young Doctor Malone.

Just one more thing.

Our complaining friend was honest. His light did flare up and his TV tube did die. Why shouldn't he conclude that we ought to buy him a new tube?

His lights flared up for two reasons. One, as you have seen, the voltage in his house did increase somewhat during the flow of fault current. Every time the breaker reclosed, fault current flowed. But he claimed they flared up and then died back down to normal. Why?

Here's why. Whenever the lights go out, the pupil of your eye dilates to allow more light to enter the eye. Then, when the lights suddenly come back on, the pupil is still dilated and the light seems much brighter than it did just before they went off. But the eye quickly adjusts, the pupil contracts, and the lights appear to dim.

We proved to our own satisfaction that our system disturbance was not the primary cause of his equipment failure. But we did pay him for the calculated remaining life of his TV tube as a public relations gesture.

Our investigation was not made with the purpose of proving our complaining friend wrong. It was made to assure our-

UNIQUE POWER SYSTEM PROBLEMS—SOLVED

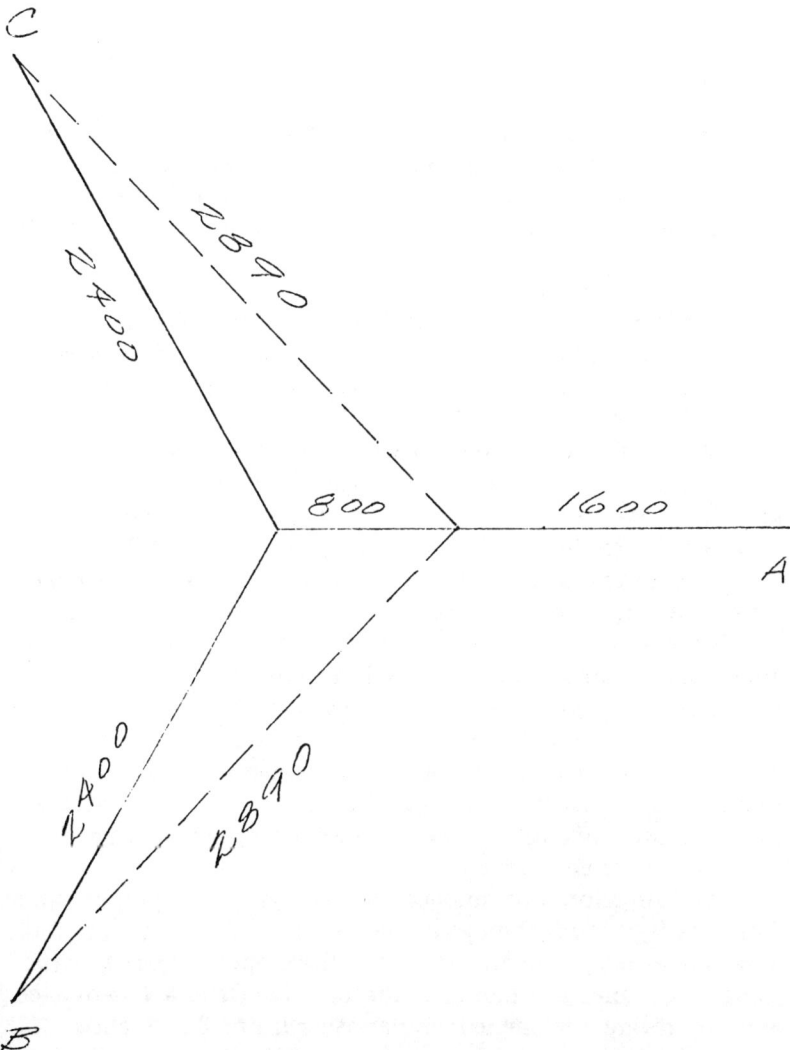

Fig. 3

selves that our system was operating in such a way that we could feel confident that our good customers were well protected from damaging or dangerous overvoltages.

9

On occasion, I ask a young engineer to explain to me how he arrived at a particular answer. His reply is usually full of, "Well, I used such-and-such chart," and, "Then I substituted my figures in such-and-such formula." Or, "I looked it up in the table."

But ask him how they got the value he found in the table or chart, or what the formula means, in plain English, and he bristles with indignation. He resents being forced to actually understand what he's doing. He is repelled by the idea of thinking, really thinking.

Much of the information we engineers use is handed to us on what amounts to tablets of stone and we are expected to take this data and use it in complete faith without ever uttering that dirty word, "Why?", or that blasphemy, "How?"

You must have had one of your arguments stopped short by the retort, "We've already run that through the computer." (Or the system analyzer.) If you continue to argue, you'll be about as popular as a bastard at a homecoming.

I am so dumb I just can't grasp anything unless I completely understand each and every step of the operation, even steps which are not really needed to "get the answer." I don't think I really have the answer until I feel I can put each element of the problem into the context of a common everyday analogy.

Let's take an example.

Transmission bus impedances are given to you in table form, derived from the system analyzer. If three lines converge at a single bus, you are told that the impedance is a certain value if all lines are closed; another value, if line 1 is opened; another, if line 2 is opened; and another if line 3 is opened.

This is all you need to work any problem you are likely to face.

However, I resent the fact that they hide from me the impedance of line 1, line 2, and line 3. Doesn't the analyzer know?

Just to maintain my self-respect I had to determine how to find out these impedance figures from the data given.

Here's the picture:

UNIQUE POWER SYSTEM PROBLEMS—SOLVED

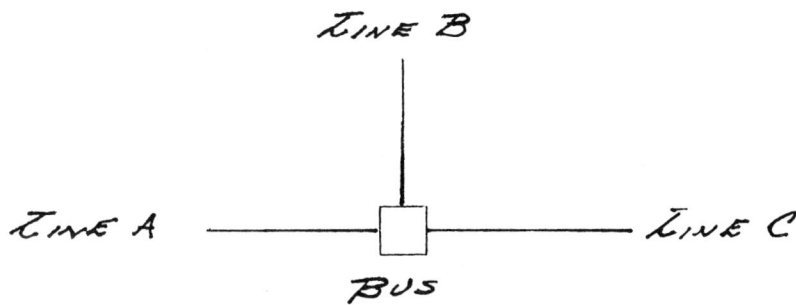

Let's call A the impedance of source A. Let's call (AB) the impedance of source A and source B in parallel; in other words, with line C open. Note that we assume that lines A, B, and C all find their way back to a common source so that (AB) can be shown thus:

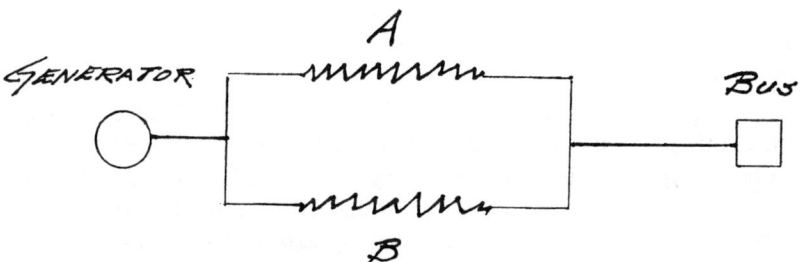

We'll call the parallel impedance of all three (ABC), etc.

Now, I will assume you do remember the simple formula for paralleling two impedances and that you fully understand why the formula is correct. Then:

If line C is open, $(AB) = \dfrac{AB}{A+B}$

If line A is open, $(BC) = \dfrac{BC}{B+C}$

If line B is open, $(AC) = \dfrac{AC}{A+C}$

$(ABC) = \dfrac{C(AB)}{C+(AB)} = \dfrac{A(CB)}{A+(CB)} = \dfrac{B(AC)}{B+(AC)}$

Solve for C:

$C(AB) = (ABC)(C + AB) = (ABC)C + (ABC)(AB)$
$C(AB) - C(ABC) = (ABC)(AB)$
$C(AB - ABC) = (ABC)(AB)$

$$C = \frac{(ABC)(AB)}{(AB) - (ABC)} \text{ or } \frac{(AB)(ABC)}{(AB) - (ABC)}$$

We now have line C impedance in terms of quantities originally given. We can deduce A and B by inspection.

$$A = \frac{(BC)(ABC)}{(BC) - (ABC)} \qquad B = \frac{(AC)(ABC)}{(AC) - (ABC)}$$

From the reciprocals of these three impedances we can readily determine the strongest, middle, and weakest source.

10

Nothing delights an engineer quite so much as to take an insignificant little project and turn it into a lifetime career. Anything that sets him to operating a calculator for hours and hours gives him a feeling of accomplishment which completely obliterates his conscience. Look how hard he's working!

The truth is that he's probably wasting his time and taking money under false pretenses. Some real thought before he starts on the problem would probably produce a method of attack which would allow him to complete his assignment in a fraction of the time. And this method would then be available to shorten the solution to many similar problems throughout many future years of problem-solving.

An Electrical Engineer will spend a large percentage of his working life calculating voltage drops and power loss. Most engineers do this by referring to tables or by using formulae which someone has worked out for him.

I have always been too lazy to even try to look this busy, so when I realized, early in my career, that voltage drop and power loss calculations were going to be my bag for the rest of my life, I started casting about for some way to simplify the process.

I reasoned that I only needed to know the percent voltage drop and power loss for 1000 kw for each 1000 feet of conductor for each size of wire. If I had this for three power factors, .8, .85, and .9, I could work most any problem that came my way by a simple calculator operation. For instance, 1500 kw for 3000 feet. 1.5 × 3.0 × my per unit figure gives the answer.

The following tables may be useful to you, too.

Table 1

4.16 kv
1000 kw/1000 ft

Copper	Voltage Drop			Power Loss		
	.8	.85	.9	.8	.85	.9
8	5.14	4.94	4.80	6.34	5.63	5.07
6	3.08	2.96	2.88	3.80	3.38	3.04
4	2.14	2.05	1.98	2.41	2.12	1.90
2	1.56	1.46	1.36	1.50	1.34	1.21
1	1.36	1.25	1.15	1.20	1.05	.945
1/0	1.19	1.07	.972	.945	.837	.751
2/0	1.05	.945	.855	.753	.677	.607
ACSR						
4	3.08	2.96	2.89	3.68	3.38	3.04
2	2.15	2.07	1.94	2.41	2.14	1.90
2/0	1.39	1.28	1.19	1.24	1.09	.973
266.8	.896	.805	.722	.593	.532	.475
397.5	.745	.669	.585	.426	.369	.335
795.0	.570	.492	.430	.209	.174	.161

For 2.4 kv 3ϕ, multiply by 3
For 2.4 kv 1ϕ, multiply by 6
For 240 V 3ϕ, multiply by 3 (but 100 kw/100 ft)

Table 2

8.32 kv

1000 kw/10,000 ft

Copper	Voltage Drop			Power Loss		
	.8	.85	.9	.8	.85	.9
8	12.8	12.3	12.0	15.9	14.1	12.7
6	7.7	7.4	7.2	9.50	8.45	7.60
4	5.35	5.12	4.95	6.02	5.30	4.75
2	3.90	3.66	3.40	3.75	3.35	3.02
1	3.40	3.13	2.88	3.00	2.62	2.36
1/0	2.98	2.68	2.43	2.36	2.09	1.88
2/0	2.62	2.36	2.14	1.88	1.69	1.52
ACSR						
4	7.70	7.40	7.23	9.20	8.45	7.60
2	5.38	5.18	4.85	6.02	5.35	4.75
2/0	3.48	3.20	2.98	3.10	2.72	2.43
266.8	2.24	2.01	1.80	1.48	1.33	1.19
397.5	1.86	1.67	1.46	1.06	0.92	0.84
795.0	1.42	1.23	1.07	0.52	0.43	0.40

Table 3

12.47 kv

1000 kw/10,000 ft

Copper	Voltage Drop			Power Loss		
	.8	.85	.9	.8	.85	.9
8	5.70	5.49	5.33	7.04	6.27	5.63
6	3.42	3.29	3.20	4.22	3.76	3.38
4	2.38	2.28	2.20	2.68	2.36	2.11
2	1.73	1.62	1.51	1.67	1.49	1.35
1	1.51	1.39	1.28	1.33	1.17	1.05
1/0	1.32	1.19	1.08	1.05	.930	.835
2/0	1.17	1.05	.950	.837	.753	.675
ACSR						
4	3.42	3.29	3.21	4.09	3.76	3.38
2	2.39	2.30	2.15	2.68	2.38	2.11
2/0	1.55	1.42	1.32	1.38	1.21	1.08
266.8	.996	.895	.802	.659	.591	.528
397.5	.828	.743	.650	.473	.410	.372

7.2 kv 3ϕ, multiplied by 3
7.2 kv 1ϕ, multiplied by 6

UNIQUE POWER SYSTEM PROBLEMS—SOLVED

Another table which I developed as a time saver gives me the weight of conductors, their current carrying capacity, their R + jX impedance per 1000 feet and per mile. It also gives me the voltage rise over these conductors per 600 Ckva installed per 10,000 feet and per mile for 12 kv and per 1000 feet and per mile for 4 kv. Note that there is very little variation regardless of wire size. A good figure to keep in mind is 3% per mile per 600 Ckva on 4 kv and .3% per mile per 600 Ckva on 12 kv.

Table 4
%V Rise/600 Kvar
Capacitors

| | Lbs | | R + jX | R + jX | %V Rise per 600 Kvar | | | |
| | | | | | 12.47 kv | | 4.16 kv | |
CU	1000 ft	Capacity	1000 ft	Mile	10,000 ft	Mile	1000 ft	Mile
8	50	90	.656 + j.161	3.460 + j.850	.623	.329	.558	2.95
6	80	120	.413 + j.157	2.180 + j.829	.607	.321	.545	2.88
4	128	170	.263 + j.150	1.390 + j.792	.580	.307	.520	2.75
2	205	230	.167 + j.145	.882 + j.765	.561	.296	.503	2.65
1	258	270	.132 + J.142	.697 + j.750	.550	.290	.493	2.60
1/0	326	310	.105 + j.140	.554 + j.739	.542	.286	.486	2.56
2/0	411	360	.083 + j.137	.438 + j.723	.530	.280	.475	2.51
3/0	518	420	.066 + j.134	.348 + j.707	.518	.274	.465	2.45
4/0	653	480	.053 + j.131	.280 + j.692	.507	.268	.454	2.40
250	772	540	.045 + j.128	.238 + j.675	.495	.261	.444	2.34
ACSR								
4	58	130	.425 + j.161	2.245 + j.850	.623	.329	.559	2.95
2	92	190	.267 + j.162	1.410 + j.855	.627	.331	.562	2.97
2/0	184	270	.134 + j.158	.707 + j.834	.611	.323	.545	2.89
266	290	460	.066 + j.124	.349 + j.655	.480	.254	.430	2.27
397	432	600	.045 + j.120	.238 + j.633	.464	.245	.416	2.20
CW								
6A	102	130	.418 + j.159	2.205 + j.840	.615	.325	.552	2.91
4A	162	180	.263 + j.154	1.390 + j.813	.596	.314	.534	2.82

11

IZ is always equal to, or greater than, IR. Right? Right. Then it seems a little wicked to find that our calculations sometimes give us a larger percentage I^2R than IZ.

"So what?" you say, "What's the difference? All I want is the answer."

Remember the statistician who drowned wading across a creek that averaged only three feet deep? He had the right answer but he didn't really understand his problem.

Solving the I^2R-IZ riddle will not provide you with a ready tool to whop other problems over the head with, but it will give you a better understanding of what voltage drop actually is and it will drive away most of the confusion which voltage calculations inevitably cause.

Let's get several things straight, first of all. Things I find to be twisted up in most young engineers' minds to a hopeless degree. And since it is a religious belief with us engineers to never admit we don't fully understand a thing, we find it difficult to get our brains unscrambled. How many times have you seen one engineer trying to get an answer while maintaining all the time that he really knows the answer already? The other engineer carefully, and condescendingly, explains a solution in detail despite the fact that he obviously doesn't understand what he is saying. The disgusting thing is to watch them part, both fully satisfied.

Percentage is one number divided by another number, multiplied by 100. It is a thing's per-unit-value on a base of 100.

Percent voltage drop does not fit this definition because our charts all assume that voltage drop is such a small portion of the sending voltage that no matter the angle of dangle of the IZ vector, V_s is always parallel to V_r. With V_r at $0°$, we assume that the horizontal component of IZ is close enough in length to the actual difference in length of V_r and V_s to produce no appreciable error. And that's true. But it does set the first trap in the path of our clear visualization of voltage drop.

UNIQUE POWER SYSTEM PROBLEMS—SOLVED

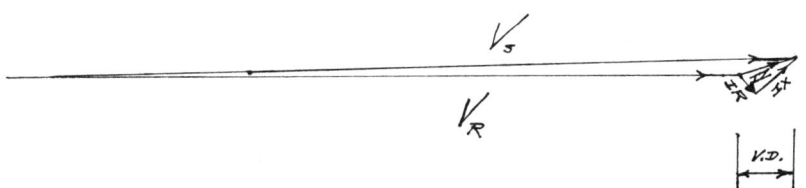

Let's draw an extreme case:

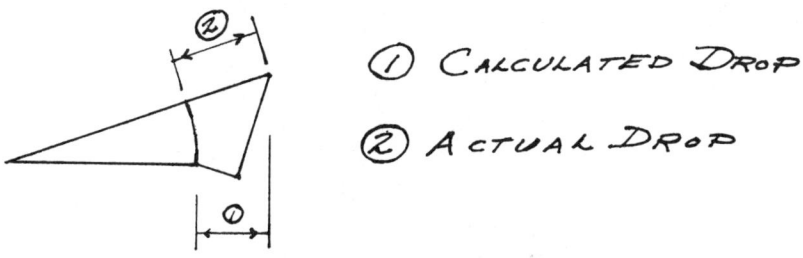

① CALCULATED DROP

② ACTUAL DROP

"2" is obviously bigger than "1."

Unfortunately, we do run upon actual field situations where the calculated voltage drop is as much as 40%. We stay in business by injecting regulators and installing capacitors, and hiding from our customers.

Note that if you use my method of drop per 1000 kw per 1000 feet and get an answer bigger than 20% drop, you probably have more than 20% drop actually. If you want to go for real accuracy you must revert to basic calculations of current, at its angle, times impedance, at its angle, and compare the scaler values of V_s and V_r.

We can solve our little problem, assuming normal voltage

drops (less than 20%), so that "voltage drop equal to the horizontal component of IZ" is valid.

$$\% \text{ loss} = 100 \times \frac{I^2 R}{IV_R \times P.F.}$$

$$= 100 \frac{IR}{V_R \times P.F.}$$

$$\% \text{ V.D.} = 100 \times \frac{IR \times P.F. + IX \sin\theta}{V_R}$$

$$\frac{\% \text{ V.D.}}{\% \text{ loss}} = 100 \times \frac{\frac{IR \times P.F. + IX \sin\theta}{V_R}}{100 \times \frac{IR}{V_R \times P.F.}}$$

$$= \frac{(IR \times P.F. + IX \sin\theta)(V_R \times P.F.)}{(V_R)(IR)}$$

$$= (P.F. + \frac{IX}{IR} \sin\theta)(\frac{V_R}{V_R} \times P.F.)$$

$$= P.F. (P.F. + \frac{X}{R} \sin\theta)$$

Now let's calculate this expression for several sizes of wire and plot the results as a family of curves.

Our first wire size will be an imaginary one which has only resistance, no reactance. Notice that the amount of resistance is of no concern to us.

	#?—no. X x/r = 0			#6 Cu x/r = .380		
(1)	(2)	(3)	(4)	(2)	(3)	(4)
P.F.	x/r sinθ	(1) + (2)	(3) × P.F.			
1.0	0	1.0	1.00	0	1.0	1.0
.95	0	.95	.90	.119	1.069	1.01
.9	0	.9	.81	.167	1.067	.96
.8	0	.8	.64	.228	1.028	.82
.7	0	.7	.49	.272	.972	.68
.6	0	.6	.36	.304	.904	.54
.5	0	.5	.25	.330	.830	.41
.4	0	.4	.16	.348	.748	.30
.3	0	.3	.09	.363	.663	.20
.2	0	.2	.04	.373	.573	.10
.1	0	.1	.01	.378	.478	.05
0	0	0	0	.380	.380	0

(1)	(2)	(3)	(4)	(2)	(3)	(4)
P.F.	x/r sinθ	(1) + (2)	(3) × P.F.			
	#8 Cu × x/r = .246			#4 Cu x/r = .570		
1.0	0	1.0	1.00	0	1.0	1.0
.95	.077	1.067	1.01	.178	1.128	1.07
.9	.108	1.008	.907	.251	1.151	1.04
.8	.147	.947	.757	.342	1.142	.92
.7	.176	.876	.613	.408	1.108	.78
.6	.196	.797	.479	.456	1.056	.63
.5	.213	.713	.356	.494	.994	.50
.4	.226	.626	.250	.523	.923	.37
.3	.235	.535	.161	.544	.844	.25
.2	.241	.441	.088	.560	.760	.15
.1	.244	.344	.034	.568	.668	.07
0	.246	.246	0	.570	.570	0
	#1 Cu x/r = 1.08			#4/0 Cu x/r = 2.84		
1.0	0	1.0	1.0	0	1.0	1.0
.95	.337	1.287	1.22	.773	1.723	1.64
.9	.476	1.376	1.24	1.25	2.15	1.94
.8	.649	1.449	1.16	1.70	2.50	2.00
.7	.773	1.473	1.03	2.03	2.73	1.91
.6	.865	1.465	.88	2.27	2.87	1.73
.5	.937	1.437	.72	2.46	2.96	1.48
.4	.991	1.391	.56	2.60	3.00	1.20
.3	1.03	1.330	.40	2.71	3.01	.90
.2	1.06	1.260	.25	2.81	3.01	.60
.1	1.075	1.175	.12	2.83	2.93	.29
0	1.08	1.08	0	2.84	2.84	0

Plotting the results, we get a graph which looks like this:

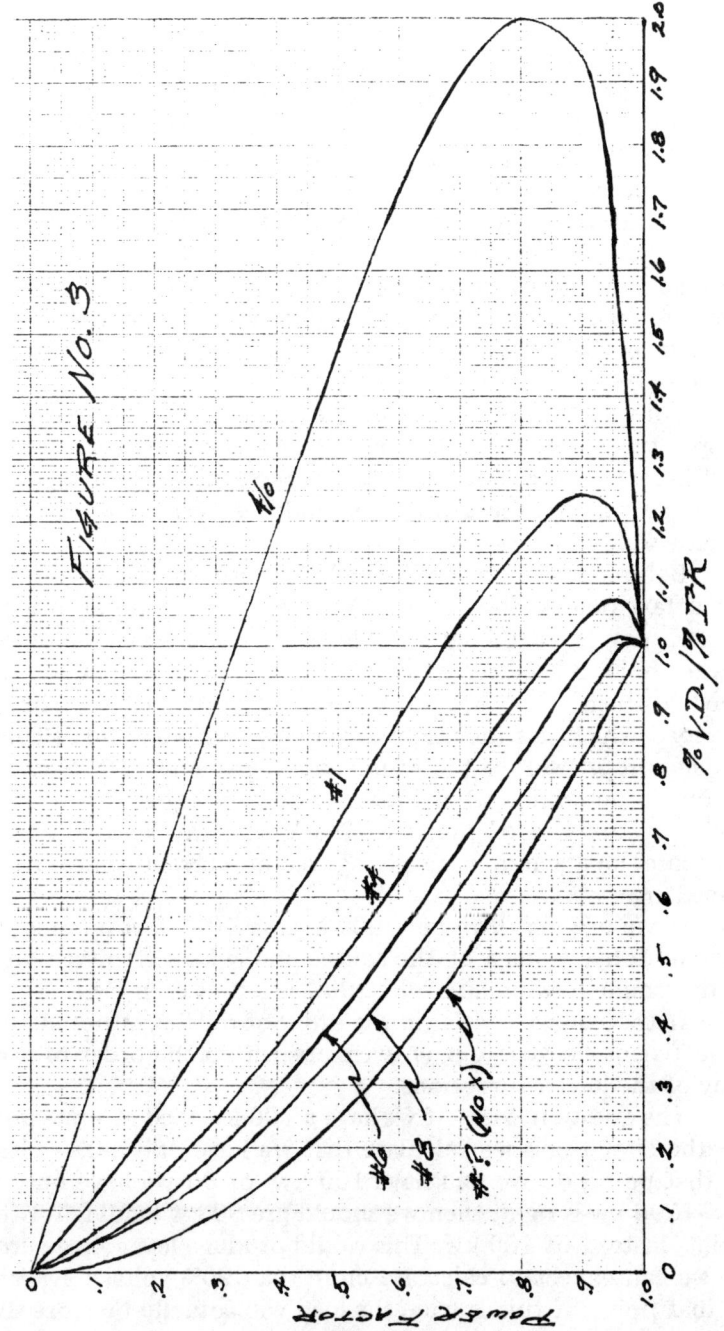

Note that the ratio of % voltage drop to % $I^2 R$ loss is ruled by two variables, impedance angle and power factor angle, with the impedance angle tending to increase the ratio and the power factor angle tending to decrease the ratio as the power factor angle increases.

Impdenace angle exceeds 45° for #1 Cu and larger and the % V.D./% $I^2 R$ ratio never goes below unity for power factors over 65%, using conductor this big or bigger. However, for small wires you must look more to losses to justify conductor replacement or conversion and less to increased load due to increased voltage.

Conversely, note that on 4/0 Cu at .8 P.F. the % voltage drop is twice the % loss. This suggests the necessity for a radically different approach to justification of a project involving existing large wire to the attack we would use when we start with small wire.

Before leaving the subject of calculating voltage drop, let's get a few points straight. First, we start by assuming loads at various points on the circuit. Next, we assume that these load figures hold, no matter what voltage we finally conclude is applied to them. Then, we assume that 100% voltage is applied to the load which has the minimum voltage available and calculate voltage drops back to the source, adding them to 100% voltage to finally determine what source voltage is required to produce 100% at the end of the line. That's the way we do it if we calculate long hand, multiplying vector current times vector impedance.

If we use my drop per 1000 kw per 1000 feet method, we assume 100% voltage at the source and subtract voltage drops to determine how much of the 100% is left at each load.

If the voltage drop is in the order of 5-10%, either method is sufficient because our answers are within the order of accuracy of the load assumptions.

The question arises, if we have a 100 kw load at some point on the circuit and we calculate that we have only 90% voltage at that point, do we still have 100 kw, or do we have only 90 kw? If 90 kw is right, then we should use 90 kw loading for that point, instead of 100 kw. This would produce less voltage drop, so we can expect to calculate more than 90% voltage available at that point. If this is true, the load will actually be more than 90 kw, and so on and on ad nauseum.

UNIQUE POWER SYSTEM PROBLEMS—SOLVED

Point load is determined by reading actual line current and adjusting it to take into account that any field reading we make is unlikely to catch the peak load, or we add kva ratings of connected transformers and multiply by an assumed diversity factor.

Obviously, assumed point loads are the result of our best educated guess. They should be those loads we would expect to have with proper voltage supplied. If voltage drop is excessive, we must offset it with regulators and capacitors to maintain a level which will keep apparatus in reasonable working order.

So you see we pick our point loads as best we can and then stick with them, even if our calculations lead to excessive drops.

Even if low voltage causes the biscuits to bake slowly, the housewife will continue to bake until they're done. This will lead to more coincidence of loads and higher peak loads.

12

There have been many lively arguments through the years among engineers over the value to place on increased load as the result of decreased voltage drop. If we increase the voltage to a particular load by one percent, how much load increase results? Then after we conclude that argument, how much is each kw of increased load (due to increased voltage) worth?

The first question causes us to look at the known effect of increased voltage on various types of loads. For instance, we know that if we increase the voltage on a resistance load 10%, the current through that load increases 10%. The load is the multiple of the voltage and the current, 110% × 110%, or 121%. So resistance loads increase by the square of the voltage.

Motor loads don't increase with increase of voltage because, as the voltage increases, the current decreases correspondingly. So the load remains constant.

But if the resistance load is an oven and the lady of the house is baking a roast, the heat is usually thermostatically controlled, so that theoretically no increase in kwh results. However, if she's been fighting a slow-cooking stove, and now she finds that it performs as it was designed to perform, she will be inclined to make more use of it.

The same is true of use of all types of equipment, including motors. I have seen it happen time after time in areas where we have corrected a low-voltage situation. People tend to go out and buy appliances they wouldn't have considered buying before.

Lighting is a big part of the load on any electric circuit. Incandescent lighting is resistance load. When the voltage increases, the load tends to increase by the square of the increase in voltage. The light intensity goes up and nobody rushes to the switch to turn off the extra light. No thermostat turns it off automatically.

From this, you can see that there is no way to precisely calculate the increase in load due to increase in voltage. But most engineers seem to agree that a one percent increase in voltage produces at least a one percent increase in demand, on peak. My observation bears this out.

How much is this one kw worth? It only exists at peak load. It is reasonable to assume that at light load the voltage will be no better after system improvement than before, unless the voltage was really lousy before. So this kw will not follow the load curve of the rest of the load. If it did, we could expect this kw to result in 1 kw × 8760 hrs × 60% load factor (or whatever your system load factor is), for an annual total of 5250 kwh. At $.02/kwh (or whatever the average low step of your rate would produce), we find an annual revenue increase of $105.00. At this point, we can only guess what the actual kwh's should be, but half of this amount is about as close as anyone can come. Since we also can expect increased usage as a result of improved service, $50.00 per year per kw of increased load due to increased voltage under the circumstances given here is a conservative estimate.

Just the sight of electric crews working in the area actually acts as a catalyst to business activity. Customers are sure we "know something they don't know," and if a man is trying to decide whether to make a business investment in the area, this seems to tip the balance.

You can see two things from this:

1. There are definite economic benefits to improving service voltage which should be considered in making a decision on investing in system improvement. We need to evaluate these

UNIQUE POWER SYSTEM PROBLEMS—SOLVED

benefits carefully in the light of local load factor, rates, and the character of the loads involved. This is a major determinant in the decision to invest or not, and how much investment is justified.

    2. There is no scientific means to make this evaluation. Experience and intuition must be relied upon.

13

A problem which separates the engineers from the clerks is the one involving calculations on a two-phase and neutral system. This is not to be confused with an honest three-phase, or a single-phase circuit. It is devious and treacherous and can lead to some extremely disappointing results.

    But it is a terrific exercise to make us see and understand the interaction of currents, impedances, and power factors. You will find that balancing the loads on a two-phase system made up of different wire sizes on phases and neutrals may be the worst thing you can do. Take a simple example (see illustration, page 46).

6 Cu $.413 + j.157 = .442 \; \underline{/+20.8°}$ Ohms/1000 ft
6A Cw $.418 + j.159 = .447 \; \underline{/+20.8°}$ Ohms/1000 ft
4 Cu $.263 + j.150 = .303 \; \underline{/+29.7°}$ Ohms/1000 ft
2 Cu $.167 + j.145 = .221 \; \underline{/+41.0°}$ Ohms/1000 ft

"A" Phase Currents

  50 kw $\underline{/-30°} = 50/7.2 =$ 6.95 a. $\underline{/-30°}$ = 6.02—j3.48
100 kw $\underline{/-30°}$ = 13.90 a. $\underline{/-30°}$ = 12.04—j6.95
200 kw $\underline{/-30°}$ = 27.80 a. $\underline{/-30°}$ = 24.08—j13.90
250 kw $\underline{/-30°}$ = 34.75 a. $\underline{/-30°}$ = 30.10—j17.40
300 kw $\underline{/-30°}$ = 41.70 a. $\underline{/-30°}$ = 36.11—j20.85
400 kw $\underline{/-30°}$ = 55.60 a. $\underline{/-30°}$ = 48.15—j27.80

46 UNIQUE POWER SYSTEM PROBLEMS—SOLVED

LOADS IN KW
ASSUME THAT
CURRENTS ALL
LAG 30°

	I 2000'	II 3000'	III 10,000'	IV 30,000'
	400↑ 2cw.	300↑ 2cw.	250↑ 2cw.	200↑ 6Acw.

(A) .442 /+41.0° .663 /+41.0° 2.21 /+41.0° 13.41 /+20.8°
    .334+j.290   .501+j.435   1.67+j1.45   12.39+j4.76

        [100]   [50]   [60]   [200]

(N) .884 /+20.8° 1.326 /+20.8° 4.42 /+20.8° 13.41 /+20.8°
    .826+j.314   1.240+j.471   4.13+j1.57   12.54+j4.76
    6cw.   6cw.   6cw.   6Acw.

        [100]   [200]   [100]

                                 100↑
                                 4cw.

(B) .606 /+29.7° .709 /+29.7° 3.03 /+29.7°
    .526+j.300   .790+j.450   2.63+j1.50
    4cw.   4cw.

    400↑   300↑
    4cw.   4cw.

UNIQUE POWER SYSTEM PROBLEMS—SOLVED

"B" Phase Currents
100 kw $\underline{/-150°}$ = 13.90 a. $\underline{/-150°}$ = −12.04 − j 6.95
200 kw $\underline{/-150°}$ = 27.80 a. $\underline{/-150°}$ = −24.08 − j13.90
300 kw $\underline{/-150°}$ = 41.70 a. $\underline{/-150°}$ = −36.11 − j20.85
400 kw $\underline{/-150°}$ = 55.60 a. $\underline{/-150°}$ = −48.15 − j27.80

Neutral Currents (Assume 1/3 through the earth.)

Section IV
2/3 × 200 kw $\underline{/-30°}$ = 18.52 a. $\underline{/-30°}$ = 16.04 − j9.26

Section III is sum of:
50 kw, "A"   6.02 − j 3.48
100 kw, "B"   − 12.04 − j 6.95
2/3 × 200 kw, "N"   16.04 − j 9.26
                  + 10.02 − j19.69 = 22.09 $\underline{/-63.03°}$
                           × 2/3 = 14.73 $\underline{/-63.03°}$
                                 = 6.68 − j13.13

Section II is sum of:
50 kw, "A"   6.02 − j 3.48
200 kw, "B"   − 24.08 − j13.90
Section III, "N"   + 6.68 − j13.13
                   − 11.38 − j30.51 = 32.56 $\underline{/\,249.5°}$
                           × 2/3 = 21.71 $\underline{/\,249.5°}$
                                 = −7.60 − j20.3

Section I is sum of:
100 kw, "A"   12.04 − j 6.95
100 kw, "B"   − 12.04 − j 6.95
Section II, "N"   − 7.60 − j20.30
                   − 7.60 − j34.20 = 35.03 $\underline{/\,257.5°}$
                           × 2/3 = 23.35 $\underline{/\,257.5°}$
                                 = −5.05 − j22.8

Voltage Drops

A I
$55.6 \ /-30° \quad \times \quad .442 \ /+41.0° \ = \ 24.6 \ /+11.0°$
$ = \ 24.1 + j\,4.7$

A II
$41.7 \ /-30° \quad \times \quad .663 \ /+41.0° \ = \ 27.6 \ /+11.0°$
$ = \ 27.1 + j\,5.2$

A III
$34.75 \ /-30° \quad \times \quad 2.21 \ /+41.0° \ = \ 76.8 \ /+11.0°$
$ = \ 75.4 + j\,14.7$

A IV
$27.8 \ /-30° \quad \times \quad 13.41 \ /+20.8° \ = \ 372.8 \ /-9.2°$
$ = \ 368.0 - j\,59.6$

B I
$55.6 \ /-150° \quad \times \quad .606 \ /+29.7° \ = \ 33.7 \ /-120.3°$
$ = \ -17.0 - j\,29.1$

B II
$41.7 \ /-150° \quad \times \quad .909 \ /+29.7° \ = \ 37.9 \ /-120.3°$
$ = \ -19.1 - j\,32.7$

B III
$13.9 \ /-150° \quad \times \quad 3.03 \ /+29.7° \ = \ 42.1 \ /-120.3°$
$ = \ -21.2 - j\,36.3$

Use only 2/3 of each neutral current.

N I
$23.35 \ /257.5° \quad \times \quad .884 \ /+20.8° \ = \ 20.64 \ /-81.7°$
$ = \ 3.0 - j\,20.4$

N II
$21.71 \ /249.5° \quad \times \quad 1.326 \ /+20.8° \ = \ 28.79 \ /-89.7°$
$ = \ 0.15 - j\,28.8$

N III
$14.73 \ /-63.03° \quad \times \quad 4.42 \ /+20.8° \ = \ 65.11 \ /-42.23°$
$ = \ 48.2 - j\,43.8$

N IV
18.52 /−30°   × 13.44 /+20.8° = 248.9 /−9.2°
                             = 245.7 − j 39.8

Assume 7200 / 0° voltage across 200 kw load at end of section IV, phase "A." Voltage across 50 kw load on phase "A" at end of section III is 7200 / 0° plus drop in A IV and N IV.

7200 / 0° = 7200.0 + j   0
             368.0 − j 59.6
             245.7 − j 39.8
            7813.7 − j 99.4   = 7814.3 /−0.73°

End of II A = End of III A + III A + III N
             7813.7 − j 99.4
             75.4 + j 14.7
             48.2 − j 43.8
            7937.3 − j128.5   = 7938.3 /−0.93°

End of I A = End of II A + II A + II N
            7937.3 − j128.5
              27.1 + j   5.3
               0.15 − j 28.8
            7964.55 − j152.0 = 7966.0 /−1.1°

Start of Phase A = End of I A + I A + I N
            7964.6 − j152.0
              24.1 + j   4.7
               3.0 − j 20.4
            7991.7 − j167.7   = 7993.4 /−1.2°

Assume 7200 /−120° voltage across 100 kw load at end of section III phase "B." Voltage across 200 kw load on phase "B" at end of section II is 7200 /−120° plus drop in B III and N III.

50 UNIQUE POWER SYSTEM PROBLEMS—SOLVED

$$7200\ /\!-\!120° = -3600.0\ -j\ 6235.4$$
$$-\ \ \ 21.2\ -j\ \ \ \ 36.3$$
$$\underline{\ \ \ \ \ 48.2\ -j\ \ \ \ 43.8}$$
$$-3573.0\ -j\ 6315.5 = 7256.2\ /\ 240.5°$$

End of I B = End of II B + II B + II N
$$-3573.0\ -j\ 6315.5$$
$$-\ \ \ \ 19.1\ -j\ \ \ \ 32.7$$
$$\underline{\ \ \ \ \ \ 0.15\ -j\ \ \ \ 28.8}$$
$$-3592.0\ -j\ 6377.0 = 7319.1\ /\ 240.6°$$

Start of phase B = End of I B + I B + I N
$$-3592.0\ -j\ 6377.0$$
$$-\ \ \ \ 17.0\ -j\ \ \ \ 29.1$$
$$\underline{\ \ \ \ \ \ \ 3.0\ -j\ \ \ \ 20.4}$$
$$-3606.0\ -j\ 6426.5 = 7369.1\ /\ 240.7°$$

The voltage drop at the end of each section is the voltage at the end of that section subtracted from the starting voltage of that phase. It is customary to relate drop to received voltage to get percentage drop. As long as you are aware of that, O.K. But be sure you realize that this is not the percentage that the voltage actually drops between the two points. That would be the drop related to the voltage at the point closer to the source. But it would be unwieldy to have to compare each section's drop with the voltage at the source side of that section. It's lots handier to compare each drop to 100% voltage. Just be sure you realize that your percent voltage drop is not a precise measure of the actual voltage decay in any section.

Now, let's look at our drops.

Drop to end of section

IVA	7993.4 − 7200.0	= 793.4 =	11.0%
IIIA	7993.4 − 7814.3	= 179.1 =	2.5%
IIA	7993.4 − 7938.3	= 55.1 =	0.1%
IA	7993.4 − 7966.0	= 27.4 =	0.04%
IIIB	7369.1 − 7200.0	= 169.1 =	2.3%
IIB	7369.1 − 7256.2	= 112.9 =	1.6%
IB	7369.1 − 7319.1	= 50.0 =	0.07%

UNIQUE POWER SYSTEM PROBLEMS—SOLVED

The losses can be calculated by multiplying each section's resistance times the square of the scaler value of the current through that section.

Note that there is only 118.1° between the two voltages at the source. This is typical of two-phase and neutral voltage calculations. You know that this can't be true. The angle between the two phases at the source is actually 120°. What actually happens is that the received voltage on phase "A" is rotated counterclockwise 1.2° to +1.2°, and phase "B" received voltage rotates clockwise .7° to −120.7°. The system is responding to the absence of phase "C" and is tending to become more of a single-phase system. This effect is very pronounced with certain combinations of impedance and power factor. What we actually have is a bastard-calf single-phase system with the neutral displaced from mid-point in the direction of phase "C."

Let's look at the example on the following page of two-phase and neutral which shows how a voltage rise may result in one phase.

Notice that in this case there is no load on phase A (or very little load). The sending voltage on phase A is less than the voltage received. This is aggravated if the neutral is much smaller than the phase wires.

About all that can be said for two-phase delivery is that it ain't as bad as single-phase, but it ain't a lot better. Don't think it's anywhere near two-thirds of a three-phase circuit.

Every two-phase and neutral problem must be calculated long-hand if accurate results are essential.

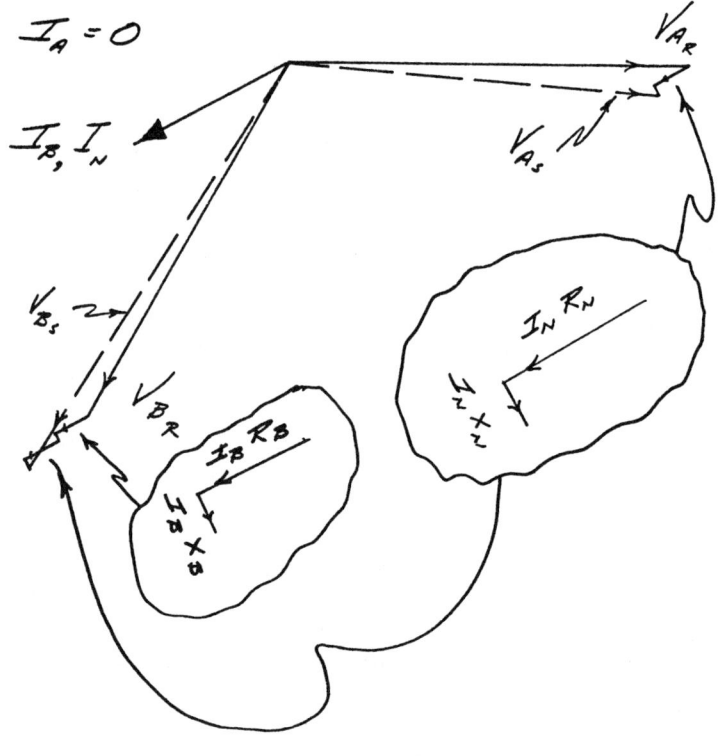

14

Little angles are to the engineer what blackheads are to the teenager. Anything under 5.75° is a pain in the neck. Anything under .575° doesn't exist. Just look at your slide rule. (What's a slide rule?)

If you don't know it already, I'll tell you. There are 360° in a circle. If a circle has a perimeter of 360 feet, its diameter will be about 115 feet. Its radius, 57½ feet. So, if I want to measure off a five-degree angle in the field, I just step off 57 feet, turn 90° and step five feet. You can measure large angles this way, too, if one of your legs is shorter than the other, so that you walk in a perfect circle.

This little trick can also be used to determine small angles when you know the horizontal and vertical sides. Divide the long horizontal side by 57½, and divide the answer into the vertical side. The answer is in degrees.

For instance, take a right triangle with horizontal side, 1100 units, and vertical side, 130 units. 1100/57½ = 19.13; 130/19.13 = 6.796°. Check on your calculator, which gives an answer of 6.4°. Close enough, right?

15

It was an old bank of three 25 Ckva capacitors connected wye on a 4 kv system. Each capacitor was rated 2.4 kv. All three units were badly damaged. They all failed at the same time.

To have one unit fail would not have raised any question. But all three at the same time! Why?

We checked and found that each unit had been fused with a 25 amp fuse link. Somebody had fused the same way they would have fused a 25 kva transformer.

Another thing we found was that the neutral point was floating. Again, just like a transformer bank would have been treated. (See diagram on the following page.)

If the neutral had been properly grounded and one unit had failed, the fault current would have passed through its fuse and out the ground. The other two units would have been unaffected. Only one fuse would have blown.

But with the neutral floating, the neutral was electrically connected to the phase feeding the failed unit. This caused full phase-to-phase voltage to be impressed across the two unfailed units.

The unfailed units blocked fault current from flowing through the failed unit, so its fuse did not blow. But with $\sqrt{3}$ voltage across each of these units, $\sqrt{3}$ times normal current passed through each undamaged unit. This was about 18 amperes, not enough to blow their 25-ampere fuses.

The two 18-ampere currents combined vectorially at the

54 UNIQUE POWER SYSTEM PROBLEMS—SOLVED

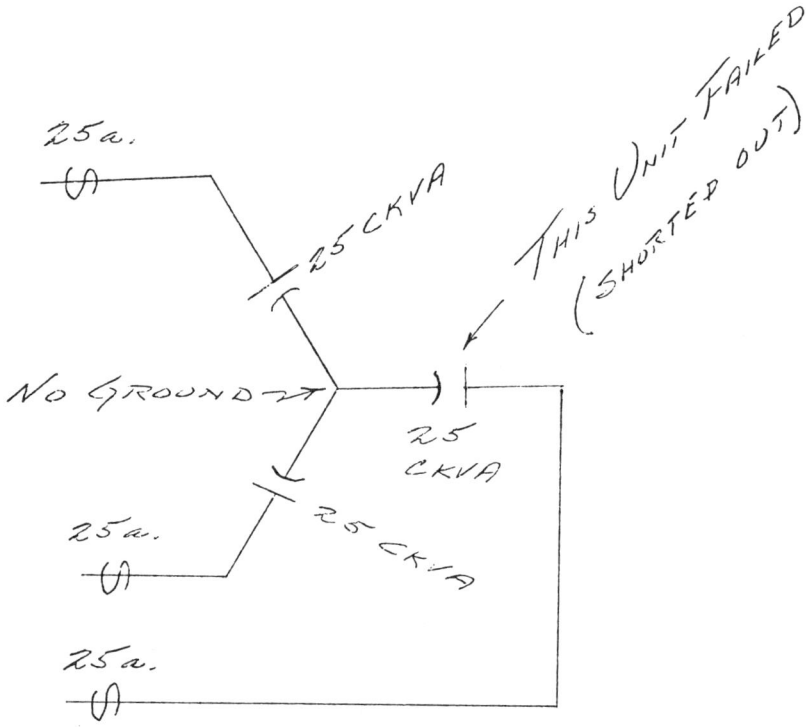

neutral point to produce about 30 amps, which passed through the failed unit on its way out that phase.

Thirty amps through that 25-amp fuse was not enough to blow it very fast. The 30 amperes persisted long enough to cause failure in the two undamaged units. As soon as they failed, full-fault current flowed, finishing off the fuse links and removing all three units from the system.

What would have happened if all units had been fused 15 amps, instead of 25 amps?

As soon as the first unit failed, the increased current of 30 amps would probably have blown the 15-ampere fuse on the failed unit before the other two units were damaged. At this point we would find the two undamaged units in series across single-phase 4 kv, phase-to-phase. The applied voltage across each unit would drop to 87% and the current would drop to 87% of normal, also. This would have left these two units undamaged.

UNIQUE POWER SYSTEM PROBLEMS—SOLVED

16

We had just completed building a new 69 kv line on "Z" structures. We energized it and, using hotstick 15 kv voltage indicators with resistors in the handles, we phased the new line out with the existing system and found to our chagrin that somebody had goofed.

We thought we had carefully walked the phases out and expected the two lines to go together with no difficulty. No such luck. It was necessary to "roll the phases." Now, as you know, that's easier said than rolled.

We looked for a convenient vertical dead-end or skeleton pole. When we found our pole, we saw that if we tried to roll the phases in just one span, the clearances would be reduced to less than acceptable wire-to-wire measurements in mid-span.

We set up models in the office using string. You couldn't walk anywhere through that office without tripping over 69 kv "phase wires." The string was too light and was affected by static electricity and drafts. We couldn't get it sagged right because the Scotch tape kept pulling loose from the doorjambs.

As a last, desperate resort, we decided to use our brains, instead of string, which was, perhaps, less than an even swap.

We realized that the wire sag had no effect at all on the relative positions of the wires (with respect to each other) at any given point in the span. We could assume that they ran from one support to the other in a straight line.

With this monumental, mind-bending discovery behind us, we then reasoned that wire-to-wire spacing at mid-span between two conductors which were not twisted but stayed in the same plane throughout the span, was the average of their spacings at the pole supports.

If the two wires were twisted 180° at one end so that their positions were reversed, they would be just in contact at mid-span, clearance—zero. Rotating 180° had reduced the mid-span clearance from unity to zero.

We noticed that if the two wires were twisted 45°, the mid-span clearance was reduced, but it was still more than half the original clearance. From this we deduced that the reduction was not directly proportional to the angle of twist.

To get a mental picture of what actually happens, let's draw a circle.

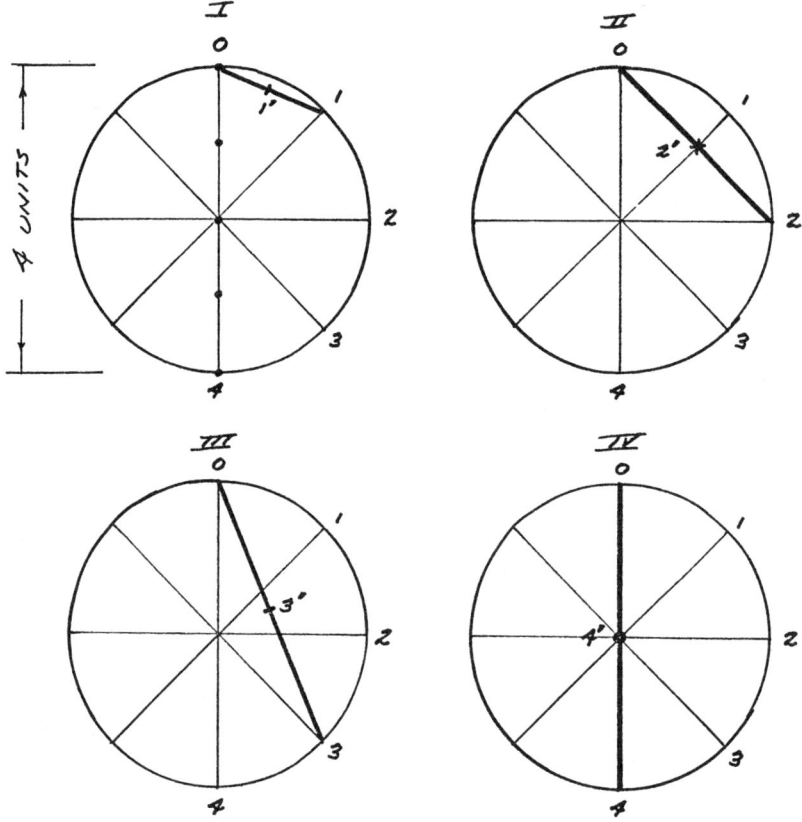

This will represent the path of the two wires which we rotate. Let's assume that we start with both pairs in the vertical position. Then we rotate one pair. First, 45°; then, 90°; then, 135°; finally, 180°. Note that while the wire end which starts at the top of the circle rotates 45°, the mid-point rotates only 22.5°. When the end rotates to 90°, the mid-point drags up to 45°, and so on.

When the wire end reaches point 1, at 45°, the wire itself is stretched between point zero and point 1. Mid-point is at the bisector of this chord at point 1'. And so forth.

UNIQUE POWER SYSTEM PROBLEMS—SOLVED

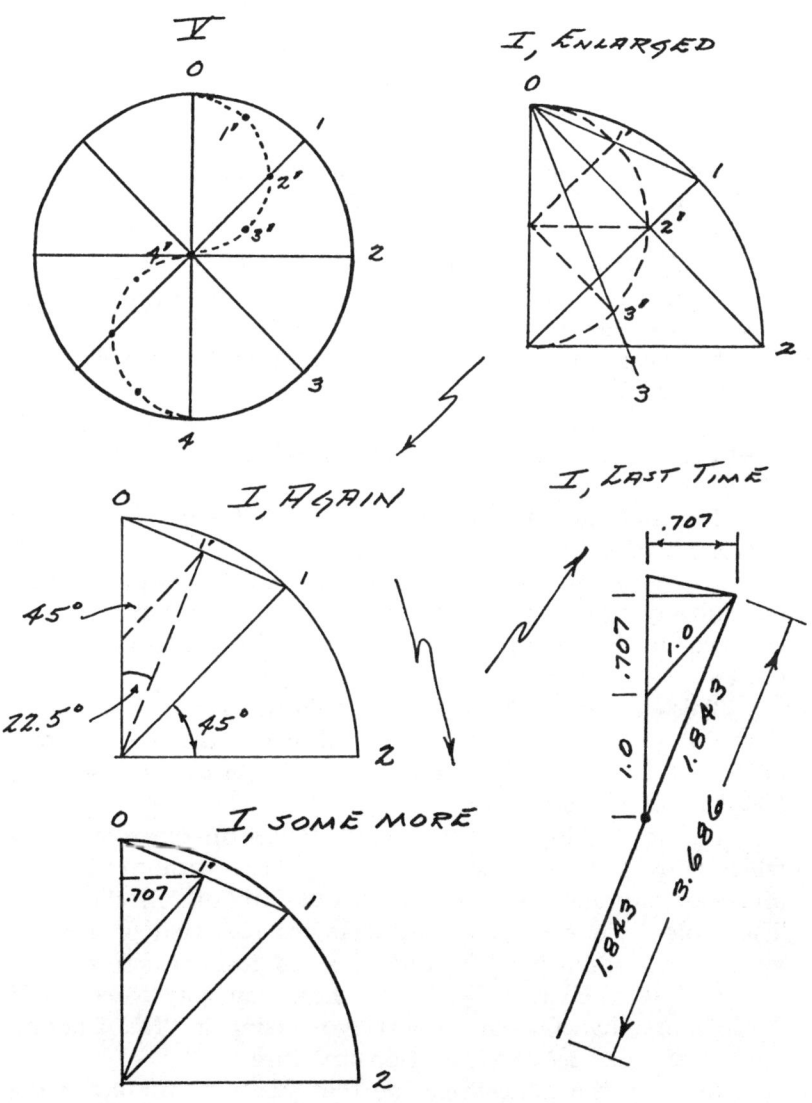

So we see that while the wire end follows the perimeter of the circle with assumed radius 2, the mid-point follows the perimeter of a circle with radius 1. The other rotating wire end and mid-point follow suit in quadrant three (not shown, to avoid confusion).

When the wire end is at 45°, the mid-point, at point 1', is .707 units east of vertical and 1.707 units north of the center of the big circle.

.707/1.707 is the tangent of 22.5° and the long side of this triangle is 1.843 units. This is half of the mid-span clearance, which then is 3.686. We started with 4 units of clearance. This has been reduced to 3.686. 3.686/4.0=.922, which is the cosine of 22.5°.

By the same process, we find that the mid-span clearance after twisting 90°, is 2.828. 2.828/4.0=.707, which is the cosine of 45°.

The mid-span clearance at 135° is 1.530 units. 1.530/4.0= .3825, which is the cosine of 67.5°.

So we see that the mid-span clearance can always be found by multiplying the average of the clearances at the two ends of the span by the cosine of half the angle by which the wire is twisted.

This suggests that the wires should never be twisted more than 60° in any one span, as that will insure that we maintain no less than 87% of full clearance. We begin to lose clearance fast if we exceed 60° twist.

I have developed some simple tables on the next pages which you can use as a guide for rolling phases without reducing mid-span clearance to less than about 90% of full clearance. The tables also avoid abruptly changing elevation of a phase which would result in a "floater" on an adjacent structure.

In Tables III and IV, I show how you may force a roll (twist more than 60° in a span) by utilizing an "H" structure (with extra spacing) as a transition structure.

In using the tables, imagine that you are standing in the line looking up at the structure shown at the top of the page. Then you back up one span, and you're now looking at the structure depicted just below the one at the top of the page. Then back up one more span, and you see the third structure down the page (if there is a third), etc.

UNIQUE POWER SYSTEM PROBLEMS—SOLVED 59

Across the top of the page are the six possible combinations. It is less confusing to number the wires 1, 2, 3, instead of A, B, C, because you may not actually know the true phase designations and if you assume wrong, there is too good a chance you'll get screwed up when you do determine the true phasing. Stick with 1, 2, 3, and you can assign true phase designations later. Of course, wire 1 must be the same phase at the start of the roll as it is at the end of the roll or you'll have gang busters. And we know to never, never, tie two wires together without first phasing them out with a meter.

The middle section of the tables shows the steps required to get to the bottom configuration in that section, with the "Z" frame turned in one direction. The bottom section shows the steps required to get to the bottom configuration in that section, with the "Z" frame turned the other way. All possibilities are covered.

60 UNIQUE POWER SYSTEM PROBLEMS—SOLVED

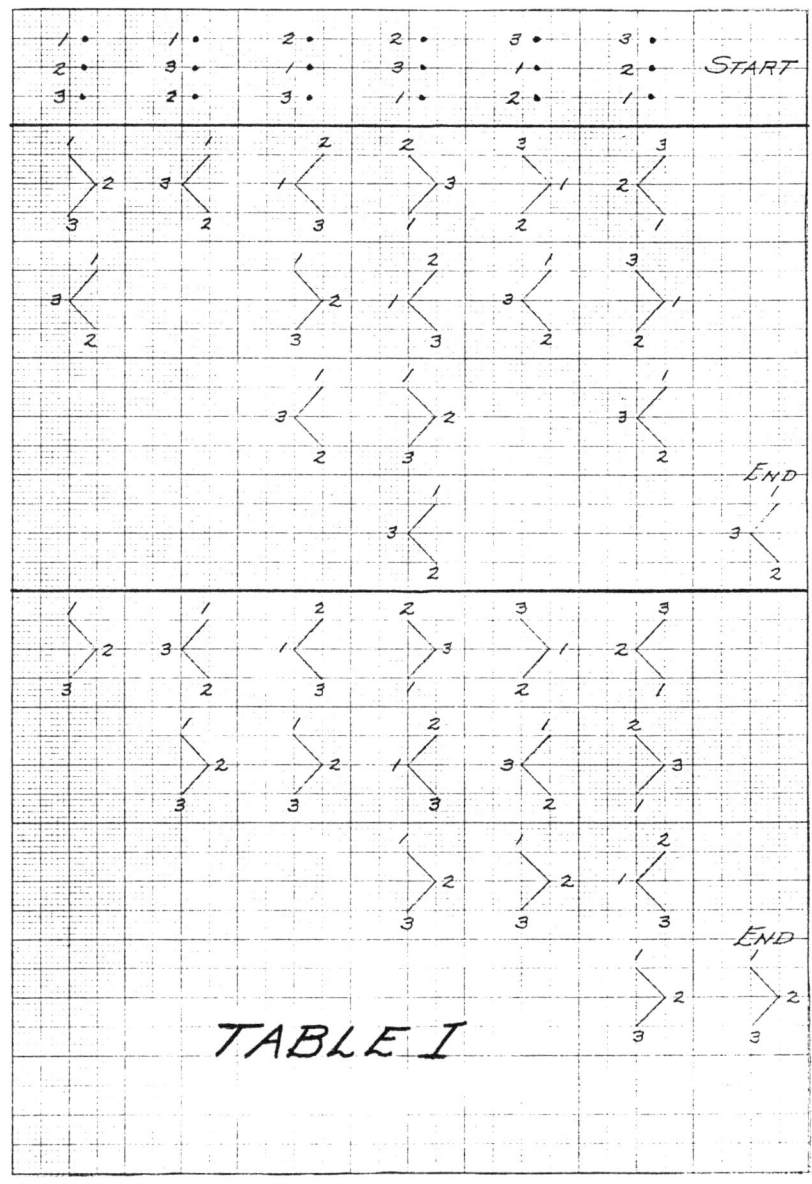

TABLE I

UNIQUE POWER SYSTEM PROBLEMS—SOLVED

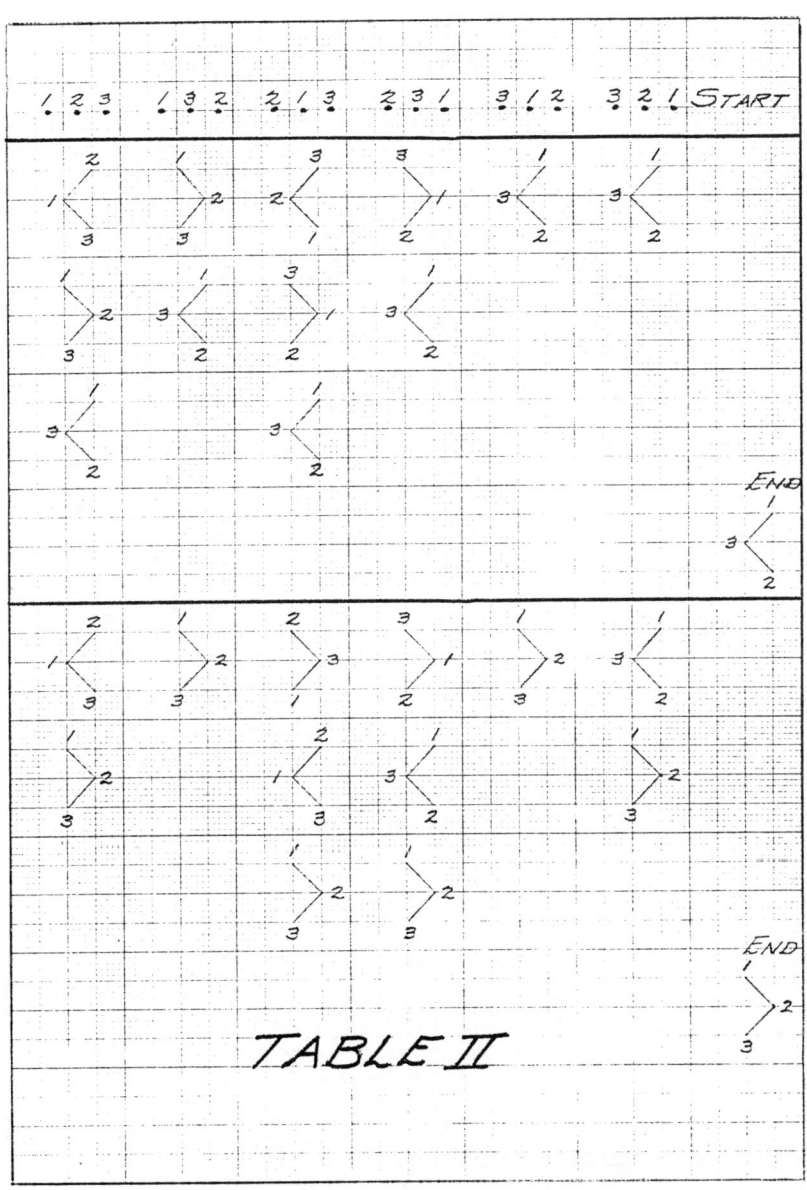

TABLE II

62 UNIQUE POWER SYSTEM PROBLEMS—SOLVED

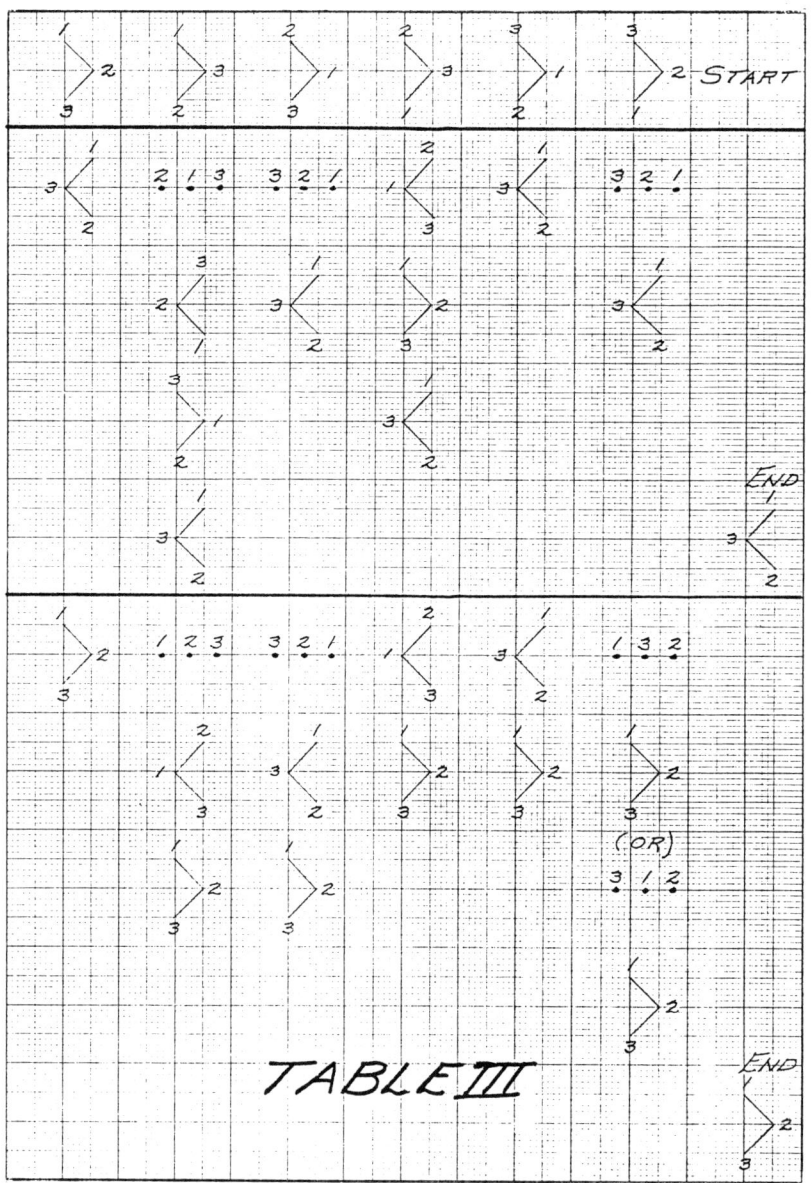

TABLE III

UNIQUE POWER SYSTEM PROBLEMS—SOLVED

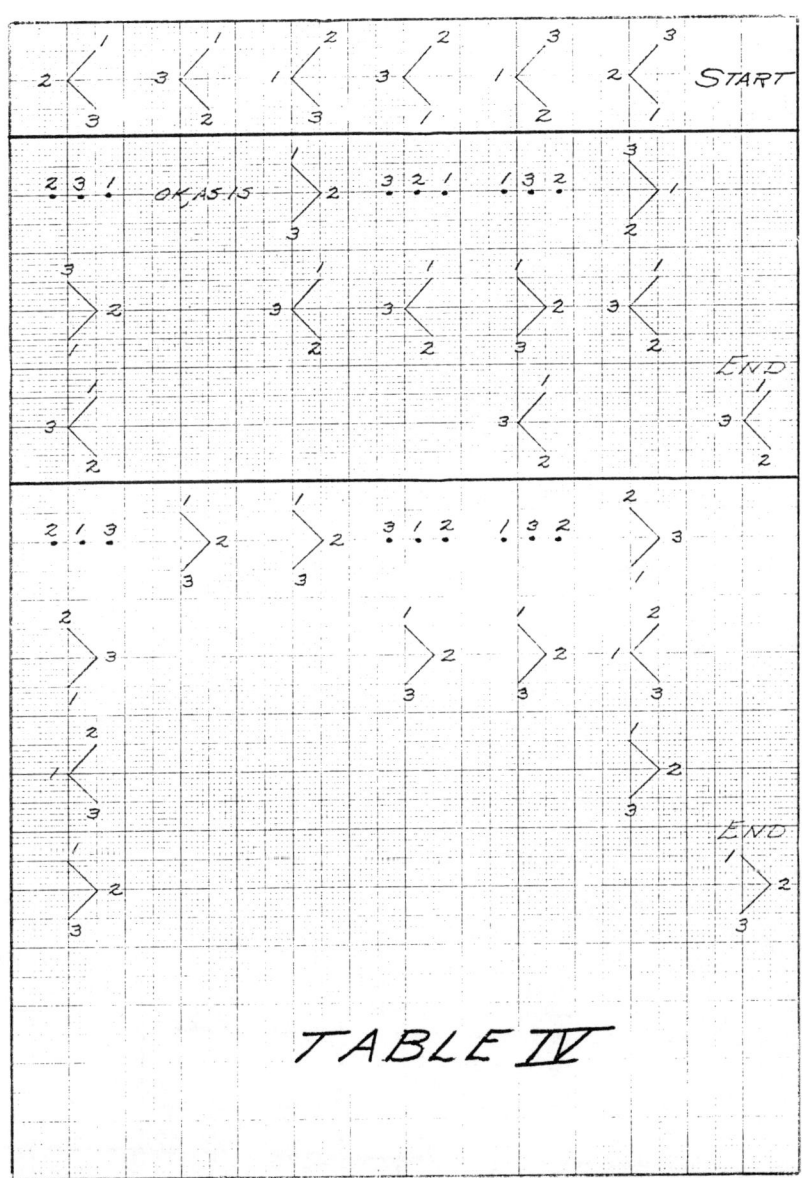

TABLE IV

17

The lights worked just fine. But when they switched on that three-phase motor, things got out of hand. It ran, alright. But not for long.

It was an open wye-open delta bank. It had just been installed. When we went out to look at it, we noticed that somehow it seemed to be neater than the usual open delta bank.

The foreman had decided it would look nicer to bring the two primary phases into the outside primary bushings of the transformers and tie the two adjacent bushings together, and to ground.

He was right. It was neat, indeed. But it made a very messy voltage diagram. The voltage across one transformer was reversed. Instead of this:

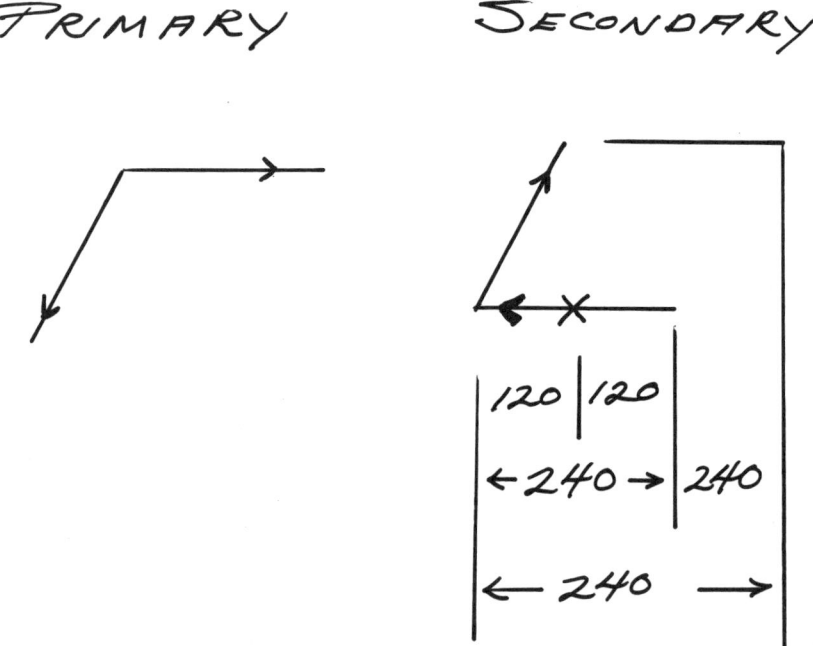

We had:

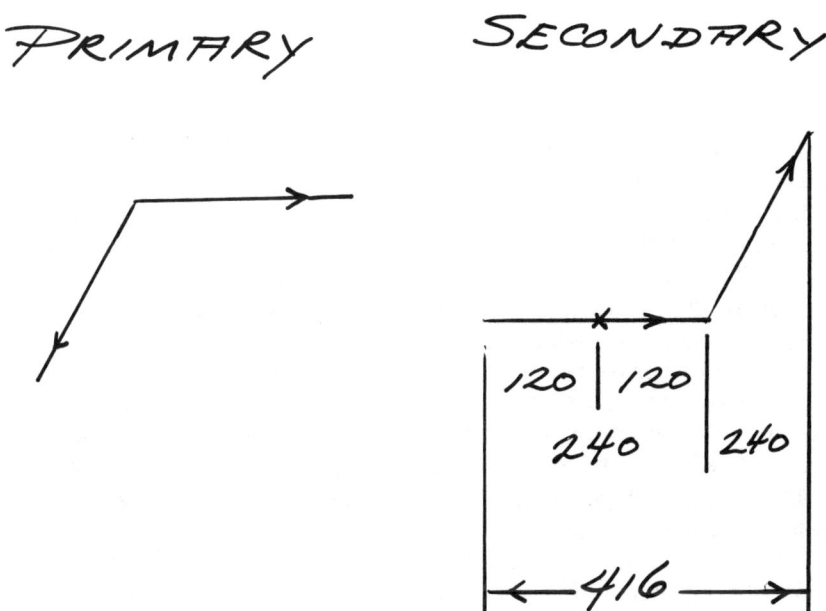

Normal voltage was impressed on all the single-phase loads, but the poor old three-phase motor was really confused. There was enough three-phase effect to get it started and keep it running, but the high voltage on one phase over-excited the iron, causing excessive heating.

You just can't explain vector relationships to linemen. For that matter, most of us engineers stay pretty confused on the subject.

To help our crew people understand the possibilities of various standard, and not-so-standard transformer hook-ups, I had them install a wye-wye bank near the crew quarters, using 240-volt secondaries. This gave us a miniature 4-kv system, divided by 10. 416 volts phase-to-phase, and 240 volts phase-to-ground. This voltage was brought into the shop area where they had three 2400/120 × 240-volt transformers mounted on a short pole, with all bushings in reach from the ground. Then they hooked those transformers up every way possible and read secondary voltages with a 50-volt voltmeter. We all learned a lot.

18

There was nothing unusual about it. Lots of times I'd been asked to supply the fault current for a new industry. "What is the capacity of the 12-kv bus from which you intend to serve our new plant?"

The fault would be controlled mainly by the size of the substation transformer and its impedance, as usual. Just get the system 69-kv impedance from the system analyzer and follow the usual steps to determine the fault current.

However, it turned out that we didn't have the system impedance at the point where the new substation was to be built.

O.K. Get the nearest system impedance, add the impedance from that point to the new sub, add the impedance of the sub, and there it is.

Only in this case the new sub was to be located at a point on the system which was fed from a regular hairnet of 69-kv lines. It looked like this, believe it or not:

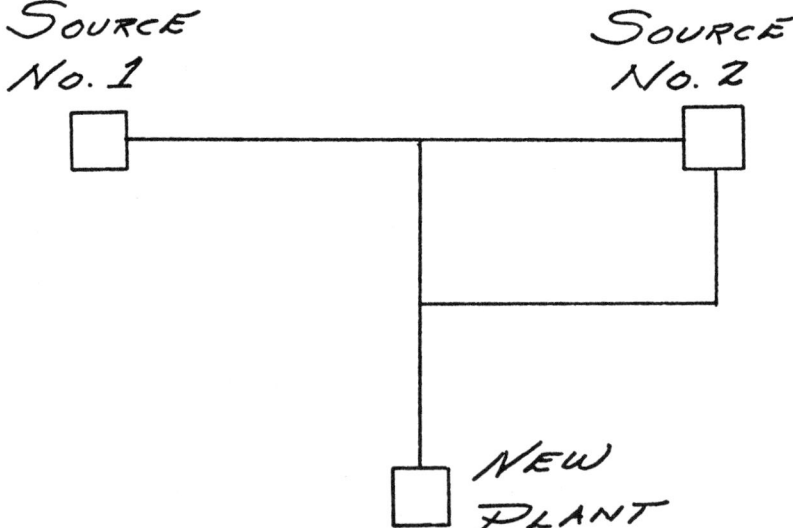

UNIQUE POWER SYSTEM PROBLEMS—SOLVED 67

From the system analyzer we had the impedance of source #1 and source #2. It was no problem to figure the impedance of each section of 69-kv line.

On a 100 Mva base the circuit impedance diagram looked like the illustration on page 68.

The transformer impedance of 7.7% on its 10 Mva base becomes $0 + j.7700$ on a 100 Mva base (100 Mva/10 Mva $\times$ .077).

When you look at this circuit and start to reduce it to a simple one consisting of one resistance and one reactance, you begin to scratch your head and you realize that it ain't all that simple.

Let's redraw it, combining all the impedances we can, without straining. The result is shown on page 69.

The upper three impedances form a wye. If we replace them with their equivalent delta it will simplify our diagram to look like that shown on page 70.

To do this we must dig back in our old textbooks and disinter the equations for converting a wye to a delta.

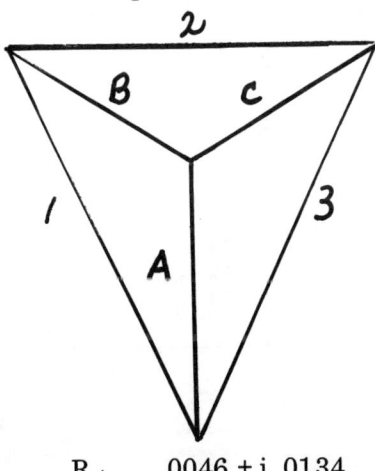

$$R_1 = R_A + R_B + R_A R_B / R_C$$
$$R_2 = R_B + R_C + R_B R_C / R_A$$
$$R_3 = R_A + R_C + R_A R_C / R_B$$

R_A	.0046 + j .0134	R_C	.0168 + j .0385
R_B	.0364 + j .1383	R_A	.0046 + j .0134
$R_A + R_B =$	.0410 + j .1517	$R_C + R_A =$	.0214 + j .0519
R_B	.0364 + j .1383		
R_C	.0168 + j .0385		
$R_B + R_C =$	.0532 + j .1768		

UNIQUE POWER SYSTEM PROBLEMS—SOLVED

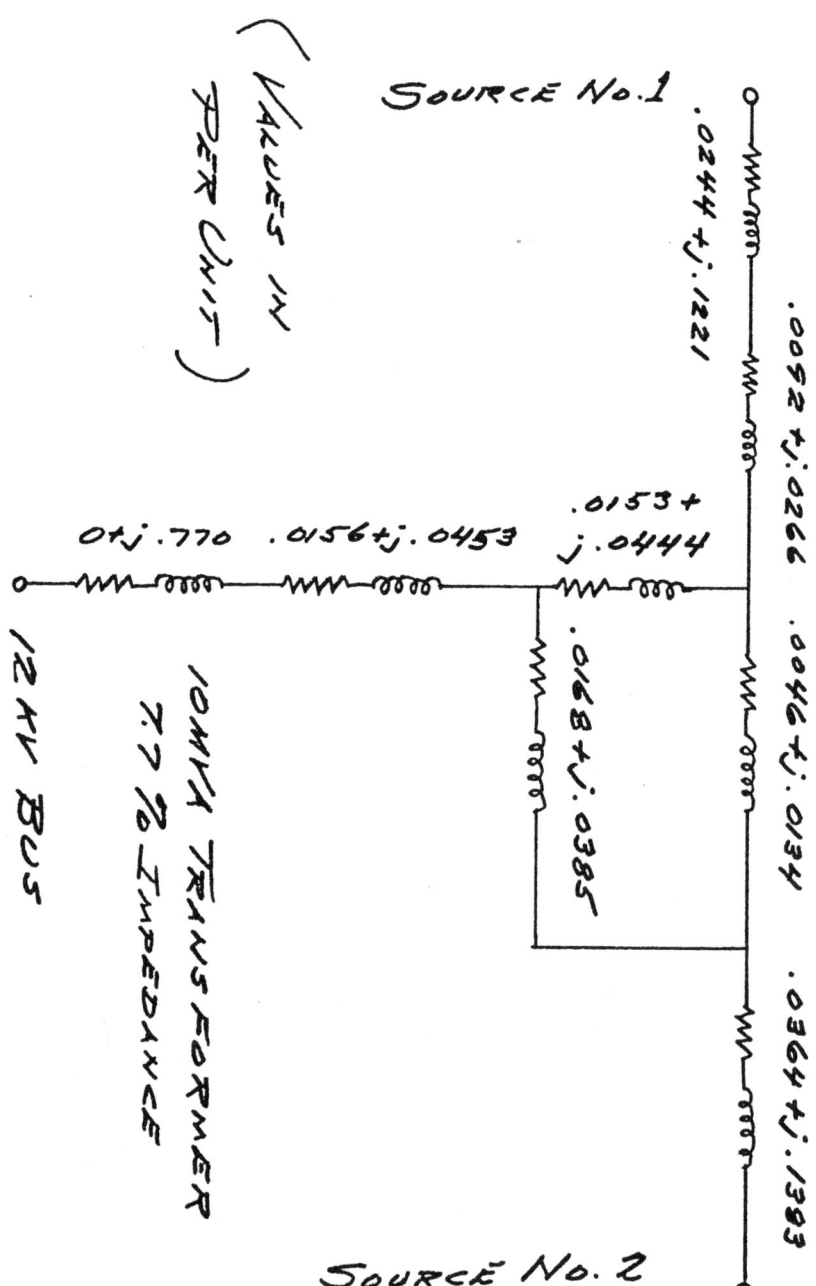

UNIQUE POWER SYSTEM PROBLEMS—SOLVED

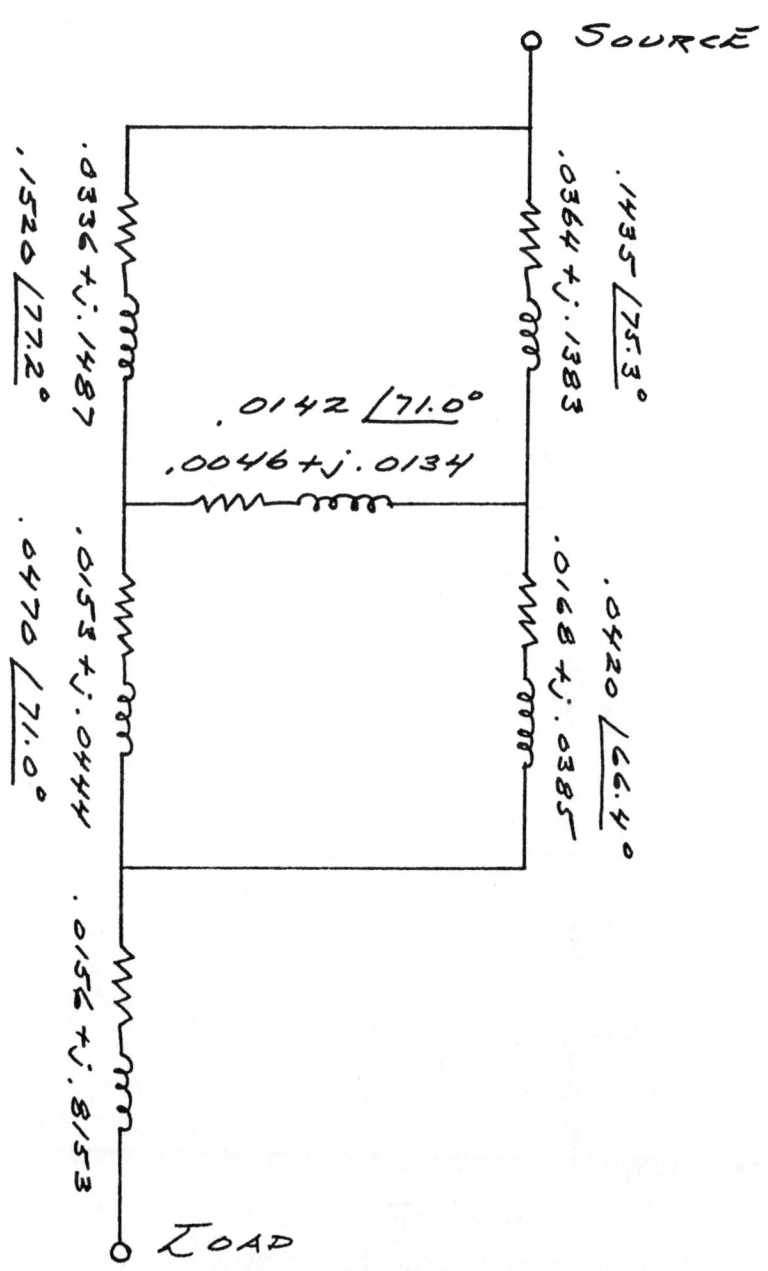

70 UNIQUE POWER SYSTEM PROBLEMS—SOLVED

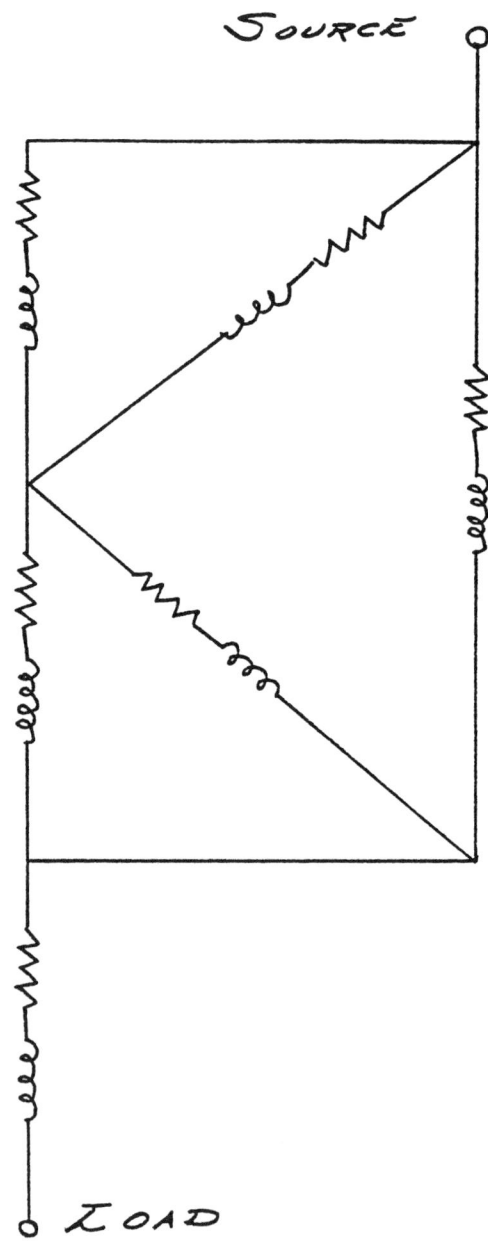

UNIQUE POWER SYSTEM PROBLEMS—SOLVED

$R_A R_B$ = .0142/71.0° × .1435/75.3° = .00204 /146.3°
$R_B R_C$ = .1435/75.3° × .0420/66.4° = .00602 /141.7°
$R_C R_A$ = .0420/66.4° × .0142/71.0° = .0006 /137.4°
$R_A R_B/R_C$ = .00204/146.3°/.0420/66.4° = .0486 /79.9°
  = .0085 + j .0479
$R_B R_C/R_A$ = .00602/141.7°/.0142/71.0° = .424 /70.7°
  = .140 + j .400
$R_C R_A/R_B$ = .0006/137.4°/.1435/75.3° = .00418 /62.1°
  = .0020 + j .0037

$R_A + R_B$ .0410 + j .1517 $R_B + R_C$ .0532 + j .1768
$R_A R_B/R_C$ .0085 + j .0479 $R_B R_C/R_A$ .1400 + j .4000
R_1 = .0495 + j .1996 R_2 = .1932 + j .5768

$R_C + R_A$ .0214 + j .0519
$R_C R_A/R_B$ .0020 + j .0037
R_3 = .0234 + j .0556

Our impedance diagram now looks like the diagram on page 72.

Combine parallel impedances:

$$\frac{.1520 /77.2° \times .2060 /76.1°}{.0831 + j .3483} = \frac{.0313 /153.3°}{.359 /76.6°}$$

.0872 /76.7° = .0201 + j .0848

$$\frac{.0470 /71.0° \times .0604 /67.2°}{.0387 + j .1000} = \frac{.00284 /138.2°}{.107 /68.8°}$$

.0265 /69.4° = .0093 + j .0248

Now we have the diagram shown on page 73.

72 UNIQUE POWER SYSTEM PROBLEMS—SOLVED

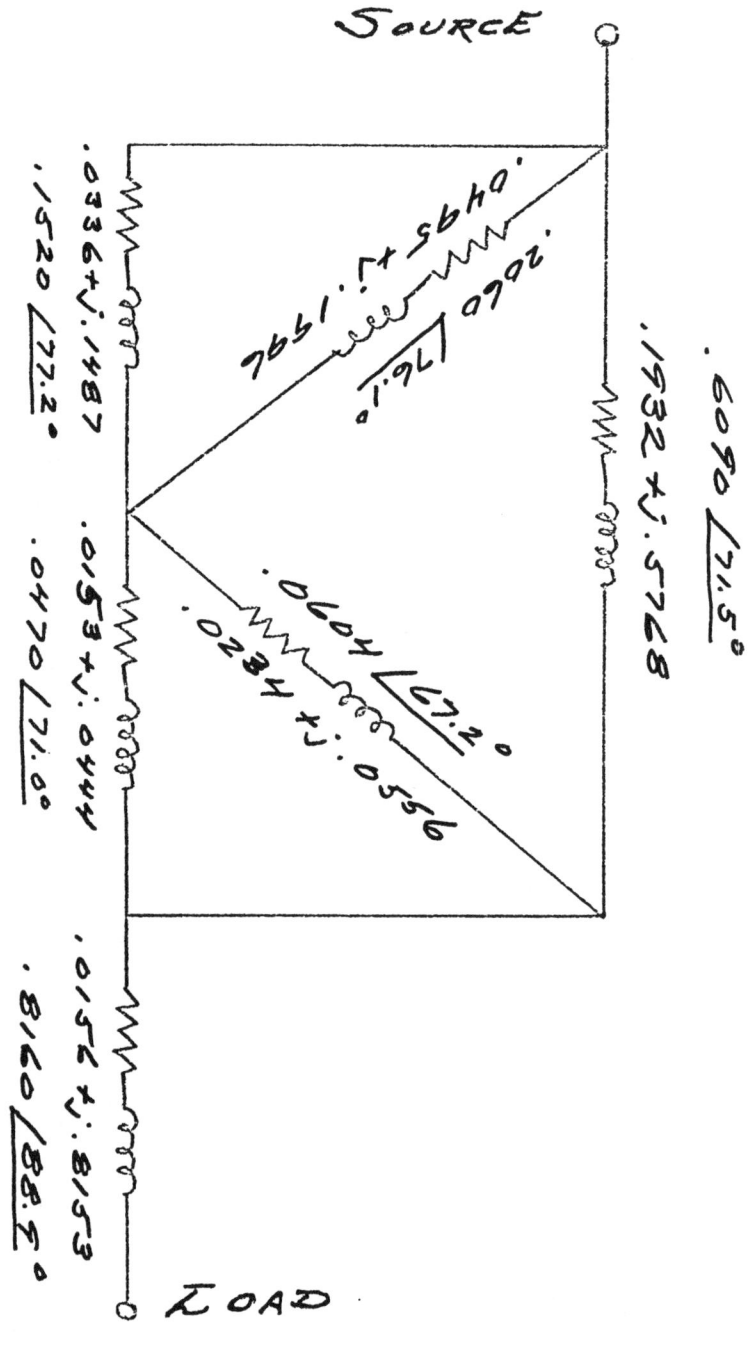

UNIQUE POWER SYSTEM PROBLEMS—SOLVED

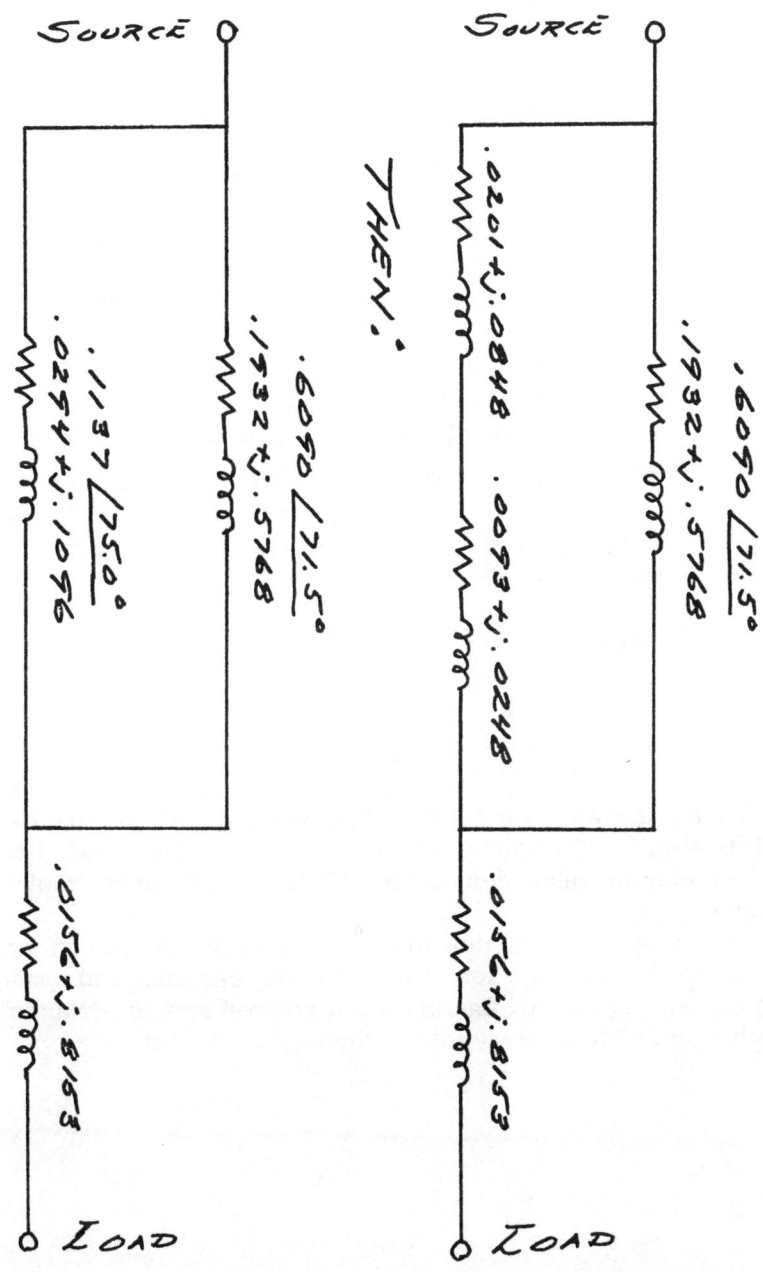

74 UNIQUE POWER SYSTEM PROBLEMS—SOLVED

Combining:

$$\frac{.6090 \underline{/71.5°} \times .1137 \underline{/75.0°}}{.2226 + j\,.6864} = \frac{.0692 \underline{/146.5°}}{.721 \underline{/72.0°}}$$

$$.0960 \underline{/74.5°} = .0256 + j\,.0925$$

plus
$$\begin{array}{r} .0156 + j\,.8153 \\ \hline .0412 + j\,.9078 \\ = .909 \underline{/87.4°} \end{array}$$

$I_{sc} = 1.0/.909 = 1.1$ per unit
$1.1 \times 4625 = 5100$ amps. (3-phase fault)
I_{sc} (phase to ground) $= 3\,E/Z_1 + Z_2 + Z_0$
$Z_1 = Z_2 = .0412 + j\,.9078 \quad Z_0 = j\,.770$

$$\begin{array}{r} .0412 + j\,.9078 \\ .0412 + j\,.9078 \\ j\,.7700 \\ \hline .0812 + j\,2.586 = 2.6 \underline{/88.2°} \end{array}$$

$I_{sc} = 3.0/2.6 = 1.153$
$1.153 \times 4625 = 5340$ amps

By the way, we got the 4625 amps by dividing the base, 100 Mva, by the voltage of the bus on which we needed the fault current value determined, 12-kv, by the square root of three.

There is nothing new in this problem. But it does offer an exercise in an almost forgotten art since computers and system analyzers. The art of figuring it out yourself and understanding what you're doing every step of the way, as you do it.

UNIQUE POWER SYSTEM PROBLEMS—SOLVED   75

19

The three-phase motors were overheating and several had been removed from the line by thermal relays. The plant manager called and wanted someone to come out and check on the situation.

The voltage read 115 — 126 — 121. The level was not too bad, but why the serious unbalance between phases?

This plant was supplied from its own substation, 69/2.4 kv delta. In the station we noticed that voltage was regulated by only two single-phase regulators. Nothing unusual about that.

In fact, we had had the third regulator removed recently for use elsewhere since these units were oversized for the load anyway.

This got us to wondering if the overheating of the motors and the regulator change-out might be connected in some way.

We sat down and drew a sketch of the normal regulator hook-up on a delta system, using three single-phase regulators.

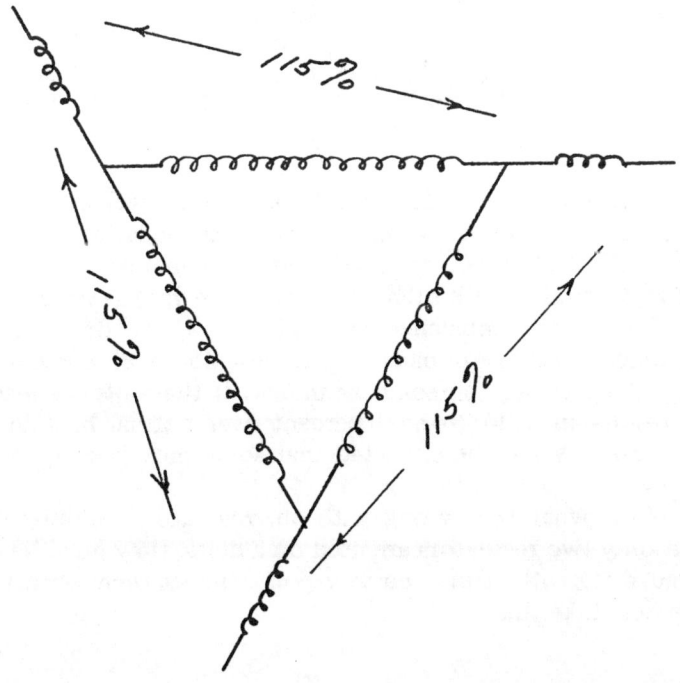

Note that with maximum raise of 10 percent, we get 15 percent raise to the system.

Now, take one regulator out:

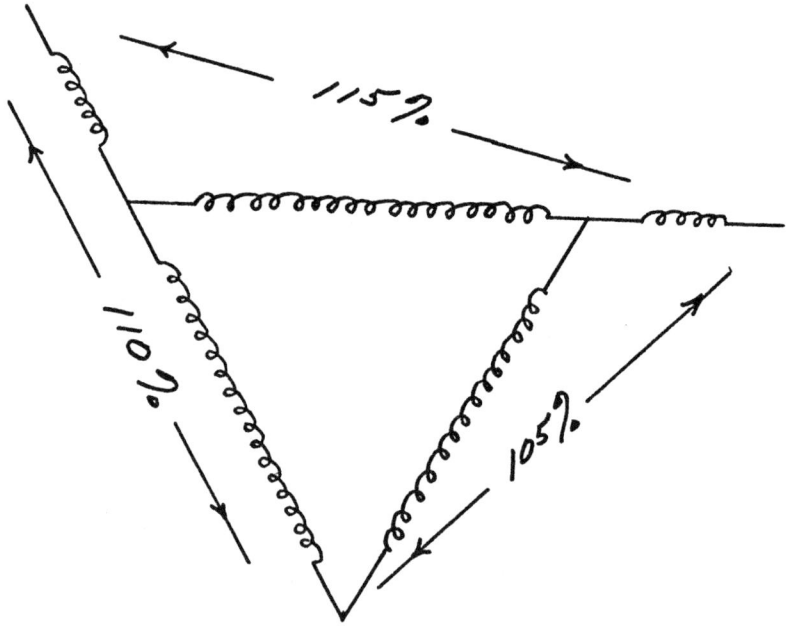

Note that with a 110-volt transmission level the primary phase voltages would be equivalent to secondaries of 115 − 126 − 121. Just what we found! And the regulators were boosting full, naturally, with such a low incoming voltage level.

The voltage unbalance was 121 − 115/121, or 5 percent. Five percent voltage unbalance produces about 25 percent current unbalance in a three-phase motor. If the motor is loaded, this results in 1.25^2, or 56 percent over normal heat in one phase. So it was to be expected that some motors dropped off the line.

Now, what was wrong with our regulator hook-up? Well, when only two regulators are used on a delta, they must be connected differently from the way you connect them with three in service. Like this:

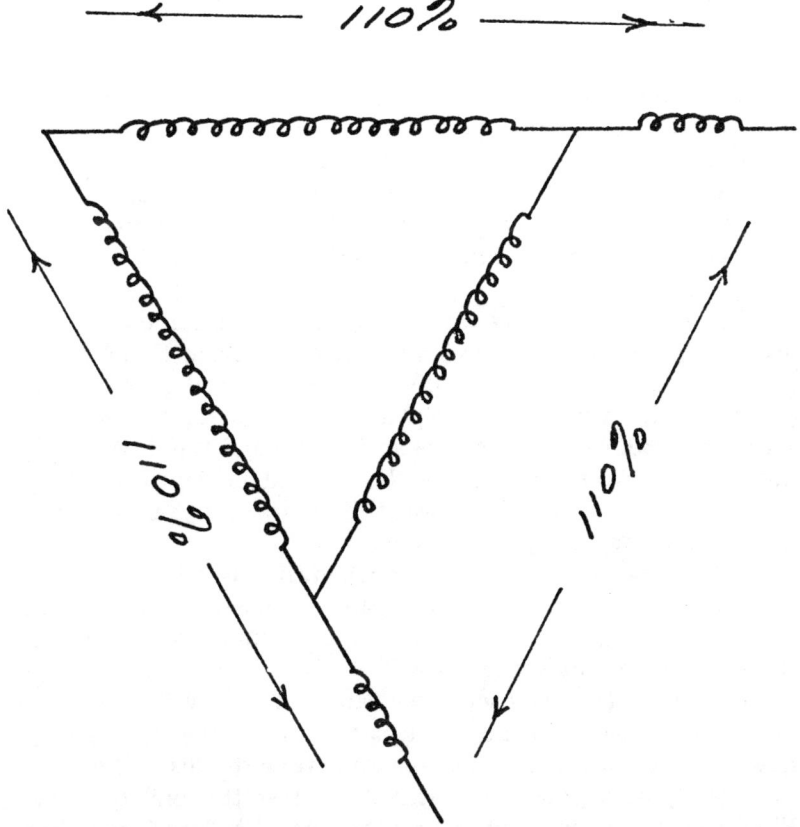

The potential coil of one regulator must be reversed.

We reconnected the regulator and rechecked the voltage. Everything was O.K. again. Everything but our explanation on the accident report. Management found it unacceptable.

20

They were afraid to reenergize the old autotransformer bank, because when the high side fuse blew on "A" phase, the lightning arrester failed on "C" phase. They wisely reasoned that some advice was in order before slamming in another fuse.

An autotransformer is a weird piece of equipment. It was not designed by engineers. It was designed by witches. Once you realize this you no longer expect it to perform like an ordinary electrical thing.

This was a 4/2.4 kv auto which had been installed at a mill years before when the town was converted from Delta 2.4 kv to WYE 4 kv. It supplied a 2300-volt motor. It had been cheaper to put up this auto than to replace the motor for the mill. Besides, the 2300-volt motor was an antique and there was every hope that it would pass away shortly and be replaced with a respectable, new 4160-volt motor. That was 20 years ago, and the old devil was still purring away.

"Was that old motor running when the fuse blew?"

"Yes, I think they tried to run a horseshoe through the mill. The old motor choked down and almost stalled. Guess the extra current blew that fuse."

"Right. Now, let's see if we can figure out why that lightning arrester gave up the ghost at the same time. By the way, how do they know it failed at the same time the fuse blew?"

"They said you could hear it all over this end of town. First, the fuse blew; then the motor started to lose speed; then they heard the L.A. pop."

Until I heard that explanation, I don't think that I had the slightest idea what caused this crazy action. But the sequence of events he described gave me a clue.

First, the motor momentarily choked down. That could pull as much as 60% of short-circuit current; plenty to blow a fuse.

Next, the lightning arrester, a 3 kv unit, failed while the motor was still running at almost full speed on single-phase power, so it was generating voltage for the third phase, keeping it hot for a few cycles.

So, at that point in time, although "A" phase was disconnected from the 4 kv source, the 2400-volt tap point was still maintained at 1400 volts away from neutral.

UNIQUE POWER SYSTEM PROBLEMS—SOLVED

However, with the autotransformer hook-up, no current flowed in the "A" phase leg of the auto. The motor current was all fed from phases "B" and "C."

The 2400-volt motor coils were connected in Delta.

When I drew the voltage diagram, the answer became apparent.

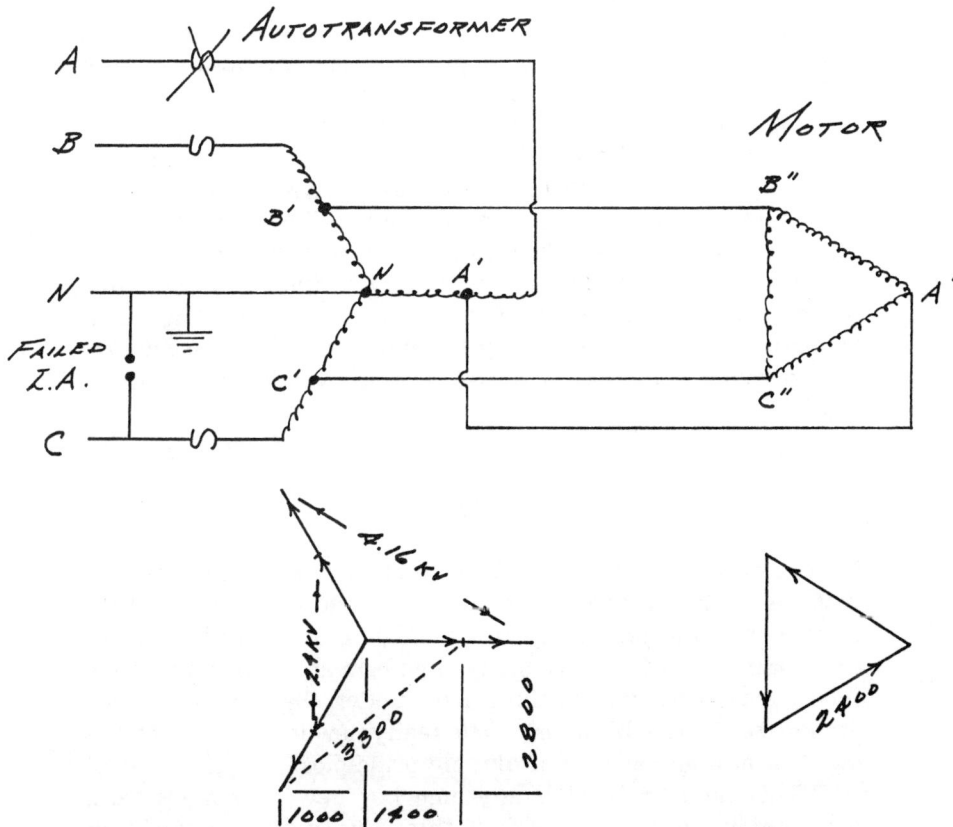

Note that when the fuse opened on "A" phase, the voltage collapsed on the autotransformer from "A" to neutral; however, the 2300-volt motor was still running, generating three-phase 2300-volt voltage. The motor Delta, A", B", C", retained its shape for some time, therefore point A' retained its relative position with respect to B' and C'. It was still 1400 volts away from neutral. Or was it? Look closely. Coil B-B'-N is energized by the source voltage from B-N. Coil C-C'-N is energized by the source voltage from C-N. But coil N-A' is not energized—no way; and point A' is maintained at its normal potential. There is no voltage drop from point A' to point N. So, lo and behold! Point N is at the same potential as point A'.

Notice that the lightning arrester that failed was hooked across C-N. Since N is now at potential A', the poor old 3-kv lightning arrester is subjected to C-A' voltage of 3330 volts.

Of course, the L.A. on phase B was similarly affected, but the operating characteristics of the motor probably resulted in enough voltage unbalance to barely save the L.A. on B while just destroying the L.A. on C. And besides, the Bible says, "One shall be taken and the other left." C was tooken.

It's that actual connection directly to the coil in an autotransformer that makes the difference. In a two-winding transformation, phase A would stay hot but there would be a difference of potential between A and N which would keep N in its place.

21

In writing this book, I am trying to stick to unusual problems that have come my way. Some of them may even seem unique to you, but I'll bet there's not a single one of my little gems that hasn't been faced by many engineers at some time during their careers. There's nothing new under the sun, and since Steinmetz, there's been very little really new in power engineering. Just new application of old, old principles.

This next problem is anything but new, and miles from being unique. But it is the sort of thing I've so often wished that somebody had put down in simple form, step-by-step, not skipping anything, so I could follow it through and clear up some of the cobwebs in my understanding of what I was trying to accomplish.

It's just a simple, little old motor-starting problem.

The fact is, it's not at all hard to work such a problem, but

UNIQUE POWER SYSTEM PROBLEMS—SOLVED

when you ask some smarty a question about it, he begins his explanation by drawing a long sigh (the length of the sigh determines his self-professed degree of superiority) and saying, "All you have to do is to simply - - -." You've spent all night Christmas Eve putting together a toy which was accompanied by instructions which invariably started off with that poison dart, "Simply - - -."

Now that you are aware of my prejudice against all snobbish know-it-alls, I'll try to make good on my promise to carry you through an example of motor starting, without leaving you somewhere along the road to hitch-hike home the best way you can.

Basically, the problem is to determine what effect the starting of a 250-horsepower motor will have on surrounding customers served from the same primary circuit, and on other equipment supplied from the same transformer bank secondaries.

Can you permit "cross-the-line" starting?

This motor was served from a 1000 kva 12/.480 kv bank (2.3% impedance), along with lots of other 480-volt equipment, including an autotransformer supplying extensive lighting.

The 12 kv primary feeder was a long one. Five miles of 397.5 MCM ACSR and then 15 miles of 2/0 ACSR conductor. The circuit was fed from a 5000 kva 69/12 kv substation. The 5 Mva unit had an impedance of 7%.

The substation was supplied from a six-mile 266.8 MCM ACSR stub tap line which started at a switching station in a main 69 kv transmission circuit. We didn't have the impedance at that station, but we did know the system impedance at a switching station 11 miles west, and at another switching station 25 miles east. See a sketch of the set-up at the top of page 82.

Let's put some impedance values on the diagram. On a 100 Mva base, one mile of 397.5 MCM 69 kv line has per unit impedance of .00526 + j.01660. One mile of 266.8 MCM 69-kv line has per unit impedance of .00780 + j.01570. 7% transformer impedance on a 5 Mva base becomes 100/5 × .07 = 1.40 per unit on a 100 Mva base.

Now we have the diagram at the bottom of page 82.

Remember that the impedance shown for station number one is the impedance shown on your analyzer printout with the line to station number two open. Likewise, the impedance

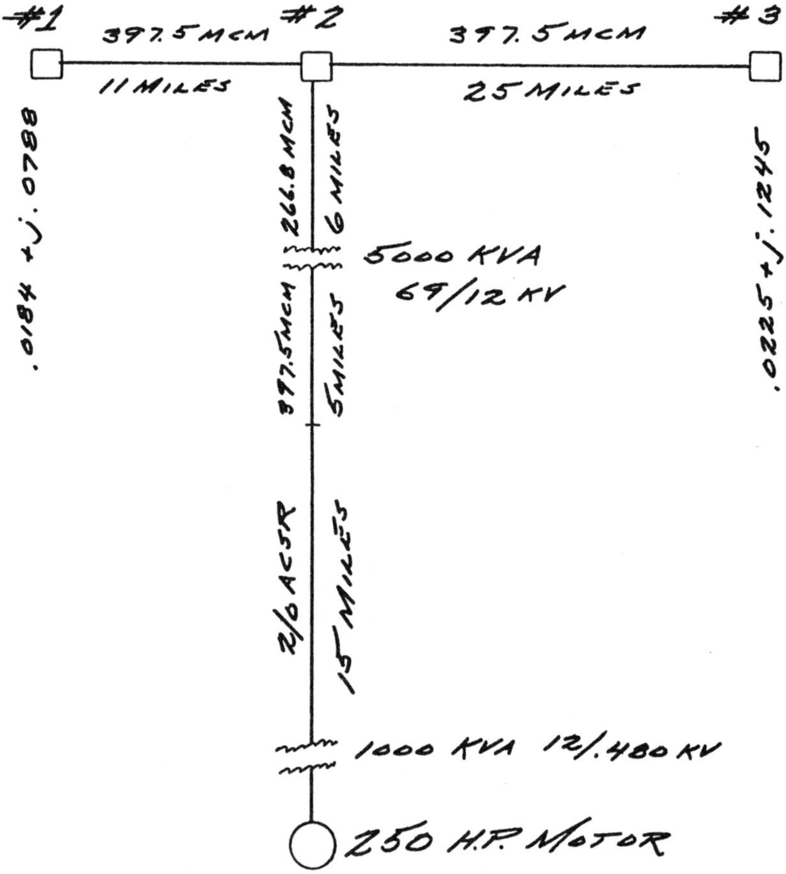

UNIQUE POWER SYSTEM PROBLEMS—SOLVED

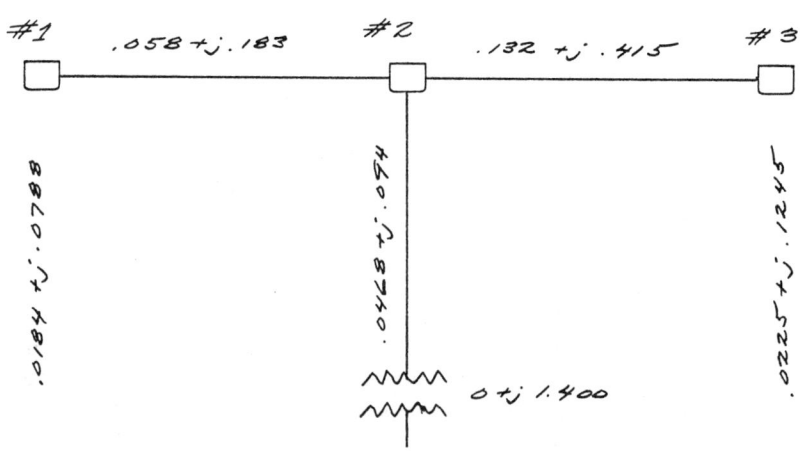

UNIQUE POWER SYSTEM PROBLEMS—SOLVED

shown for station number three is the impedance from your printout with line to station number two open. These two impedances are the "little black box" impedances at those points, looking back into the system toward the generation. All system generation can be thought of as one generator, so the two station impedances are parallel impedances closed through lines 1-2 and 3-2 at point 2, as shown in the diagram on page 84.

Now, we parallel impedances from generator to station number two. $Z = AB/A+B$.

$$\begin{aligned}&.0184 + j.0788\\ &\underline{.0580 + j.1830}\\ &.0764 + j.2618 = .274 \underline{/73.8°} = A\end{aligned}$$

$$\begin{aligned}&.0225 + j.1245\\ &\underline{.1320 + j.4150}\\ &.1545 + j.5395 = .560 \underline{/74.0°} = B\end{aligned}$$

$$\begin{aligned}&.0764 + j.2618\\ &\underline{.1545 + j.5395}\\ &.2309 + j.8013 = .832 \underline{/73.9°} = A + B\end{aligned}$$

$AB/A + B = .1535 \underline{/147.8°} \, / \, .832 \underline{/73.9°}$
$ = .184 \underline{/73.9°} = .051 + j.1768$

Impedance to point 2	$.051 + j.1768$
Impedance point 2 to sub	$.047 + j.0940$
Impedance of 5 Mva sub	$\underline{0 \;\;+ j 1.4000}$
	$.098 - j 1.6708$
$=$	$1.671 \underline{/86.64°}$

Now, let's shift over to a 12 Mva base, which is a little better for 12 kv line calculations, because it gives us figures which remain closer to unity.

Our 12 kv bus impedance then becomes $12/100 \times 1.671 \underline{/86.64°} = .2005 \underline{/86.64°} = .0118 + j.200$.

Now let's look at our 12 kv line impedances. One mile of 397.5 MCM 12 kv line has a per unit impedance of $.02 + j.05$. One mile of 2/0 ACSR is $.07 + j.07$. Five miles of 397.5 MCM, then, is $.100 + j.250$. Fifteen miles of 2/0 ACSR is $1.05 + j 1.05$. Total impedance comes to $1.15 + j 1.30$.

84 UNIQUE POWER SYSTEM PROBLEMS—SOLVED

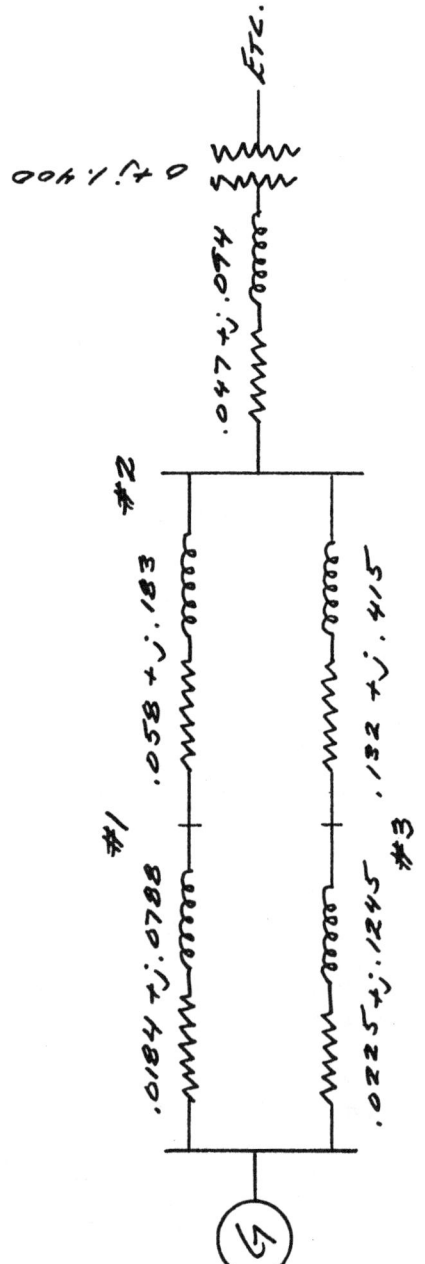

Total system impedance to the high side of our 1000 kva bank:

```
 .0118 + j  .20        12 kv bus impedance
1.1500 + j 1.30        12 kv line impedances
1.162  + j 1.50   =   1.90 /52.2°
```

This is the impedance which will control the amount of voltage dip which will be seen by customers served from the 12 kv system in the vicinity of our 1000 kva bank.

Note that if we had ignored the system impedance on the high side of the 5 Mva substation, it would have made no appreciable difference in the result. In this case, with such a long 12 kv line, even the substation impedance is of minor importance.

Let's determine our final system impedance; that of the 1000 kva bank. It had 2.3% impedance on a 1 Mva base, so its impedance on a 12 Mva base is $12/1 \times .023 = j\,.276$.

```
1.162 + j 1.500        End of 12 kv circuit
        j 0.276        1000 kva 12/.48 bank
1.162 + j 1.776        2.12 /56.8°  480 volt bus
```

We must now determine the impedance of the 250-horsepower motor at the moment of starting it. This impedance can sometimes be obtained from the nameplate or from the manufacturer, but most of the time we have to use typical values. That's what we shall do here.

When we apply voltage to a motor which is at rest, there is no back emf to oppose the applied voltage. At that instant we have a simple Ohm's Law relationship of voltage applied to a static impedance; in this case, a bunch of copper wound around a hunk of iron. This impedance is obviously mostly reactance, with a little resistance. We know from experience that the impedance angle is usually about 75° and that the current which flows upon the application of 100% voltage (which we obviously will not have at the end of a 20-mile-long circuit) will be about six times normal full-load current. From this information we can determine the impedance which will oppose whatever voltage is actually available. In fact, we shall just add this impedance to all the other impedances we have calculated and apply unity voltage to the whole system back at the generator and observe what current actually flows. In other words, we shall calculate the short-circuit current at this point with a fault impedance equal to the "standing still" impedance of the motor.

We can assume that the 250-horsepower motor takes 250 kva at full load, because the power factor will likely be in the neighborhood of .746, the ratio of kw to horsepower. So, on start, the motor with full voltage applied, will be 12,000 kva/1500 kva, or 8.0 per unit.

8.0 /75° =	2.070 + j 7.730
480 volt bus impedance	1.162 + j 1.776
Total, including motor	3.232 + j 9.506 = 10.0/71.2°

The current which will surge through the motor and be added to the system, all the way back to the generator is, then:

$$I_{sc} = \frac{E}{Z} = 1.0\underline{/0°}/10.0\underline{/71.2°} = .1\underline{/-71.2°}$$

On the 12 kv system, this will be:

$$.1 \times \frac{12 \text{ Mva}}{1.732 \times 12.47 \text{ kv}} = .1 \times 556a. = 55.6 \text{ amperes.}$$

This is somewhat less, as we expected, than the nameplate starting current would have been. It would have been about 1500 kva/12.47 × 1.732 = 69.4 amps.

Now it is a simple matter to determine the voltage dip at any point on the system by multiplying the per-unit starting current times the per-unit impedance to that point.

For instance, the dip at the 480-volt terminals of the 1000 kva transformer is:

$$.1\underline{/-71.2°} \times 2.12\underline{/56.8°} = .212\underline{/-14.4°}$$

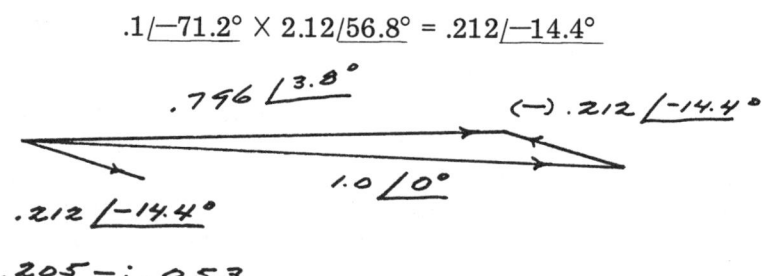

= .205 − j .053

Remember, we applied unity voltage at 0° at the generator, not at the motor terminals. So we must subtract the voltage drop to get the voltage received. $1.0 - .205 + j\,.053 = .795 + j\,.053 = .796\,\underline{/3.8°}$.

Only 79.6% voltage gets to the motor terminals. The dip is 20.4%.

This is enough to get the motor started, although it will probably grind up to speed slower than we would like. But the lights supplied from this 480-volt bank through that transformer will just plain go out. Now, if the motor is started only once or twice a day, we may get by with that. A 250-horsepower motor is not likely to be started too often, so this may not be too objectionable.

However, if this motor is on some sort of grinding, or crushing type operation, it may get choked down several times an hour. This can pull as much as 60% of starting current, so we might expect dips of about 12% as a result. This could be a real nuisance, as 12% would cause severe light blinking.

It looks like we are on the ragged edge of acceptable performance, but since the 480-volt loads are all the customer's, and it's his 250-horsepower motor that's cutting up, he may find it in his heart to put up with the inconvenience.

One other problem resulting from this severe dipping may be that other 480-volt motors will drop off the line from sensitive undervoltage relay operation. If so, the customer may have to install time-delay devices on them to hold them in until the 250-horsepower motor can get up some speed.

What about our other customers who live nearby and don't have any reason to be tolerant of bouncing voltage? The voltage dip on the 12-kv system nearby will be $.1\,\underline{/-71.2°} \times 1.90\,\underline{/52.2°} = .19\,\underline{/-19.0°}$. The received voltage is $1.0 - .179 + j\,.062 = .821 + j\,.062 = .823\,\underline{/4.3°}$.

Only 82.3% voltage is received, so the dip is 17.7%. If you permitted this, you would probably get run out of town. This much dip would cause all kinds of malfunctions of customer equipment.

It would just be asking too much of your customers to tolerate this much dip even once or twice a day. No, we're going to have to insist that a reduced voltage starter be installed on the 250-horsepower motor. This should reduce the nameplate starting current from six times full-load current to about 2½ times full-load current and our dips will be only about 40% of our calculated values; 8% on the 480-volt bus, and 7% on the end of the 12-kv circuit. This isn't the greatest, but it will probably be bearable.

22

To determine the optimum tariff, or rate, for a customer, it is often necessary to determine the value of transformer losses.

If a customer takes service at more than one point in the same vicinity, it may be to his advantage to combine his loads and take metering at one point.

However, most utilities and commissions require that the customer own all facilities past the metering point. In order to bask in the sunshine of one-point metering, the customer may be required to fork over enough dough to install or purchase one or more transformers.

As long as the meter is on the secondary side of the transformation, it does not see transformer losses. Usually, the rate associated with secondary delivery is higher than the same rate if metering is at the higher voltage, primary or transmission. This increment is usually just about enough to cover the losses and the fixed charges on the transformer investment.

In other words, the customer is not likely to save money by taking service at a higher voltage. He saves only when he combines two or more deliveries into one-point metering. That often necessitates going from low-voltage metering to metering at a higher voltage, and that brings on ownership of the transformation by the customer.

It is at this point that we must include in our rate analysis the value, or cost, of those transformer losses.

How do we evaluate these losses?

We must find out how much the customer's billing will be as the result of running the losses through his meter; and that depends on his rate. Most customers involved in one-point metering calculations are large users on some kind of power rate, as opposed to residential or commercial rates. Power rates usually recognize kwh used, kw demand, and power factor on peak. You will have to apply your rates to your calculations.

We see that we must determine the amount of demand at peak which transformer losses contribute to the total. We must estimate the number of kwh during the month by which the losses will increase the energy measured by the customer's meter. And we must determine the effect losses will have on the power factor at peak.

Then we must evaluate these three elements.

We can find the core and copper losses in manufacturer's information; however, you will find that from 167-kva to 500-kva, total full-load losses average about one percent of the transformer rating, and the ratio of copper losses at full-load to core losses is about four to one.

Core losses remain constant 24 hours a day, but copper losses vary as the square of the load. So, at half-load, core and copper losses are about equal. If the transformer is to carry a reasonable overload, as it should, at peak, the losses will begin to mount.

You can see how important it is to establish the peak loading of the transformers and the anticipated load factor before attempting to evaluate losses.

The demand charge will apply directly to the total losses at peak load because the peak losses and the peak customer load are coincident.

The effect of the losses on total power factor is seen to be negligible because they are at unity power factor. The total losses being in the order of one to two percent of the customer's peak load, we can ignore any possibility that losses will affect power factor billing.

The energy consumed by the losses is determined by multiplying the peak loss kw by the loss factor, which is a function of the customer's load factor. Since losses vary as the square of the load, it is apparent that loss factor is less than load factor for any load factor less than 100%.

Let's make up a rate, remembering that to evaluate your losses you must use your own rates.

Demand Charge
  $1.80/kw/mo — Secondary metering
  1.60/kw/mo — Primary metering
  1.45/kw/mo — Transmission metering

UNIQUE POWER SYSTEM PROBLEMS—SOLVED

Energy Charge
$0.0250/kwh/mo, First 3000
0.0150/kwh/mo, Next 7000
0.0110/kwh/mo, Next 90,000
0.0100/kwh/mo, Next 400,000
0.0090/kwh/mo, Next 500,000
0.0080/kwh/mo, Next 1,000,000
0.0060/kwh/mo, All over 2,000,000

Let's say that a customer is now taking secondary service at two points with separate billing for each service. One delivery is supplied by a 1000 kva bank with a peak loading of 1400 kw and a load factor of 80% and power factor of 80%. The other delivery is supplied by a 500-kva bank with a peak loading of 750 kw and a load factor of 50% and power factor of 75%.

One delivery is 480 volts and the other is 120/208 volts, so they can't be combined on one bank.

The customer wants an analysis to determine if he should purchase both banks and take one-point primary metering.

The 1400 kw load uses 1400 × 720 × 80% kwh/month, or 806,000 kwh/month.

Billing:
```
   1,400 × $1.80 = $ 2,520.00
   3,000 ×   .025 =      75.00
   7,000 ×   .015 =     105.00
  90,000 ×   .011 =     990.00
 400,000 ×   .010 =   4,000.00
 306,000 ×   .009 =   2,754.00
                    $10,444.00
```

The 750 kw load uses 750 × 720 × 50% kwh/month, or 270,000 kwh/month.

Billing:
```
     750 × $1.80 = $ 1,350.00
   3,000 ×   .025 =      75.00
   7,000 ×   .015 =     105.00
  90,000 ×   .011 =     990.00
 170,000 ×   .010 =   1,700.00
                    $ 4,220.00
Total  .............  $14,664.00
```

UNIQUE POWER SYSTEM PROBLEMS—SOLVED

Now combine the two on one meter. Assume that there will be noncoincidence of peaks to the extent that the combined demand will be reduced 5%. Actually, this would have to be determined by studying demand charts run on each load. (Since one load has 80% load factor, our 5% assumption is probably high.)

Billing:

```
  2,150 × .95 × $1.60 = $ 3,270.00
  3,000 × .025              75.00
  7,000 × .015             105.00
 90,000 × .011             990.00
400,000 × .010           4,000.00
500,000 × .009           4,500.00
 76,000 × .008             608.00
                       $13,548.00

                       $14,664.00
                        13,548.00
Saving . . . . . . . . . . . . . . . . . . . . . . . 1,116/mo
                            12 mo
Saving . . . . . . . . . . . . . . . . . $13,392.00/year
```

Now this saving must be balanced against the cost of purchasing the two power banks and the cost of the losses which are now metered.

The 1500 kva of transformer capacity and appurtenances can be valued at about $10/kva, or a total of $15,000. The carrying charges, including maintenance, will amount to about 20%, or $3,000/yr, $250/mo.

Now if the losses cost the customer less than $1,116 − $250 = $866/mo, it will be to his advantage to put up the $15,000 and go with one-point metering.

Let's see what the losses are worth:

First, how much loss do we have?

The 1,000 kva bank will have about 10 kw total loss at full load, of which 8 kw will be copper losses and 2 kw will be core loss. However, this bank is loaded (140/80)% at peak so the peak loss is $1.75^2 \times 8$ kw = 24.5 kw. The loss factor corresponding to 80% load factor is about 70%, so the copper losses will consume .70 × 720 × 24.5 = 12,300 kwh/mo. The core losses consume 720 × 2 = 1440/kw/mo. Total, 13,740 kwh/mo.

The 500 kva bank will have about 5 kw total loss at full load of which 4 kw will be copper losses and 1 kw will be core loss. This bank is loaded 150% with a 75% power factor at peak, so the peak loss is $(150/75)^2 \times 4$ kw = 16 kw. The loss factor corresponding to 50% load factor is about 35%, so the copper losses will consume $.35 \times 720 \times 16 = 4040$ kwh/mo. The core losses consume $720 \times 1 = 720$ kwh/mo. Total, 4760 kwh/mo.

Total for both banks is 18,500 kwh/mo.

The demand increase caused by losses will be $(24.5 + 16) .95 = 38.5$ kw.

Looking back at the combined billing for the customer's load, we see that the losses will be billed as $1.60/kw and $.008/kwh.

```
    38.5 kw  X  $1.60  =  $ 61.60
 18,500 kwh  X    .008 =    148.00
                          $209.60
```

We had $866/mo savings, not allowing for losses, so we now find that the customer will save over $650/mo, or $7,800/yr by going to one-point metering.

It won't always work out to be to the customer's advantage, depending on many factors. Loss evaluation is one of the most important of the factors in making such a determination.

23

Did you ever pour slop into a long hog trough? It kind of piles up under the stream out of the bucket and when the pile gets too high, it flows away from the middle of the trough (where you're doing the pouring) in both directions to the ends of the trough. Each little flow does its own piling up and releasing so that the slop flows away in little waves, just like dropping a stone into a pond. Except this pond is confined by the sides of the trough, so the slop is directional. When the first wave of swill gets to the end of the trough it sloshes up on the end of the trough. As it piles up against the end of the trough it builds up weight and next thing you know it starts to flow toward the center of the trough. Now it has to fight its way upstream through the delicious tidbits of garbage and slush going in the

UNIQUE POWER SYSTEM PROBLEMS—SOLVED

other direction. The process continues until you stop pouring slop. It takes a little while to settle down after you quit. By the way, I'm assuming there are no other hogs present. They wouldn't even let the slop get to the end of the trough the first time.

If you had a trough with a bend in it, say a 30° angle, twenty feet away from you toward the end, the slop would try to climb the outside wall of the trough at that point, just like a bobsled rounding a curve.

If you forgot to nail a board across the end of the trough, you would lose your slop. It would just flow right out the end and make a mess on the ground.

If you had a hole in the bottom of the trough right at the point you were doing your pouring, and the hole was just the size of your stream of slop, all the slop would go through the hole onto the ground and flow all over your feet and they wouldn't let you in the house for a long time. But no slop would flow down the trough. No waves, no splashing at an angle in the trough, no splashing at the end of the trough, no flowing out the open end of the trough. No flowing in both directions, no flowing in either direction.

If the hole in the trough were just big enough to pass half the slop you were pouring, only half of the slop would pile up and flow away in waves, splashing at angles and trough ends, or flowing out open ends.

If the hole in the bottom of the trough is fitted with a pipe just the size of the hole and the trough is mounted on stilts, say twenty feet above ground, and the pipe extends almost to the ground with a strainer fitted onto the lower end of the pipe, the slop will flow down the pipe, through the strainer and out onto the ground. If the strainer is clogged up so that no slop gets out, the slop will just fill up the pipe quickly and you might as well not have a hole in the bottom of your trough. The slop will build up in a pile over the hole and flow in waves through the end.

If the strainer is removed altogether allowing free escape for our slop, we'll be in the same shape we were when all we had was a hole with no pipe or strainer attached.

Of course, the strainer may just be partially clogged, allowing half the slop to escape freely. The slop will still back up in the pipe, but only half of the slop, so the pile of slop that builds up will be only half as big, the waves of slop flowing away only half as high, and the splashing at angles and trough ends only half as violent.

It happens to be a fact that most respectable slop-pourers pour a stream which will easily pass through any pipe which we would be inclined to use, but just to complete the picture, fancy that same idiot using a tiny little old pipe. Of course, this would act just like the stopped-up strainer and result in a slop traffic jam.

By now you electrical types will be aware that you have just had a parable laid on you.

Let's identify the elements of the parable:
    Slop — Lightning
    Trough — Conductor
    Trough-end — Dead-end
    Angle — Point of Intermediate Impedance
    Open End — Infinite Length Line
    Hole — Connection to Ground
    Pipe — Ground Wire Down Structure
    Strainer — Earth Resistance
    Open atmosphere past the strainer — Water Table, Zero Ground Resistance

We can't stop the lightning. Prayer might help, but you'll probably find that the answer to your prayer, "Please stop the lightning," will be a resounding, "Don't want to!"

If we can't stop the lightning, then we must give it some place to go where it won't create so much commotion and damage. That place is a great big old ball call "Earth," which is so big it can absorb any amount of electrical charge which the biggest, meanest bolt of lightning can inject into it. It's probably ejecting an equal charge in the midst of a storm halfway round the world. The Earth, she can take it.

The problems we experience from lightning are largely due to clogged-up "strainers." We say the earth resistance is too high. Actually, the Earth itself presents almost no resistance. It's the *connection* to that zero resistance mass that presents the resistance. The water table is in contact with the Earth over such a large area that it gives us a zero resistance connection to zero resistance earth. If we can get a low resistance connection to the water table, we've got it made. Unfortunately, this is usually pretty hard to do because water table is often pretty deep under the surface, and in developed areas we find that it is falling deeper into the ground all the time during the dry months.

This idea of connecting to water table to get a low resistance ground is borne out by your experience that if you get a drizzle, or light, soaking rain for a half hour or so before the fireworks break loose, you will not have nearly so much fuse-blowing and lightning damage as you get when a front moves in after dry weather, with lots of lightning right in the rain. Then, brother, call out the troops!

The high earth resistance we talk about is really a volume of earth at the base of the subject structure which has poor current-carrying qualities. A pole set in solid limestone is not likely to have a decent ground. Some clays and sand can be very high resistance, especially when dry.

A pole-butt ground is often set in a cup of rock with little or no contact with the earth. A driven ground may be in loose soil so that it makes poor contact with earth. Our problem is to make good contact with zero resistance earth, and that's probably "down there" unless you're in a swamp.

I once had a pole with a lousy ground. It was located in shaley rock on the bank of a creek. No amount of ground-rod driving did any good. I had the men run a wire from the pole ground in a shallow trench to the creek only a few feet away and immediately the ground resistance dropped to almost zero. Creeks, of course, are in contact with water table, somewhere or other.

It is time now for me to stop and assure you that I do not claim to be a lightning expert. (Neither do lightning experts.) I can only hope to present the action of lightning in terms which may be a little easier to grasp than some of the explanations you've heard. I do know some things about how it acts, based on long experience. I do know what has worked for me in trying to reduce the harmful effects of lightning. I do know that you don't fight it. Just like a mean mother-in-law, you learn to cooperate with it and always give it its way. After you have backed away from a confrontation with your mother-in-law, groveled, apologized and reasoned, all to no avail, you know what happens. Pow! Just the same with lightning. It wants in the worst way to get into the arms of Mother Earth, and if you stand in its way or try to slow it down, Pow!

Power engineers face two types of problems in trying to pacify lightning: the unshielded line, and the shielded line.

The shielded line puts a grounded wire above the phase conductors in the hope that lightning will prefer to strike it rather than sneak around it and get at the phase conductors first. Most of the time it works just that way.

When lightning strikes the shield wire and begins pouring "slop" into the trough (onto the wire), the potential piles up at that spot and the nearest hole in the trough is at the closest structure where there is a connection to a down wire to earth. As the pile grows taller, it begins to push waves of potential out in both directions. If a wave hits a dead-end, it splashes up against it to twice its original height. If it hits an angle or a discontinuity less severe than a dead-end, it sloshes up somewhat, but never as much as twice its original height. The first wave to hit a "hole," drops through the hole, down the ground wire to ground. And here is where our problems may begin.

If the ground resistance at that point is too high, the lightning current, which may be in the range of thousands of amperes, passes through that resistance and Herr Ohm takes over. The current times the resistance produces a voltage drop which may raise the potential of the ground wire going up the structure to a level with respect to the phase conductors (which are oscillating above and below ground sixty times a second) that the intervening air, or in the case of steel towers, the intervening insulation, breaks down and an arc is established. The arc is composed of ozone, which is conducting, and the power current rushes through this arc in the opposite direction to ground. Then the circuit breakers on the line must go to work to clear the fault. By that time, the lightning will have come and gone.

To prevent this ugly sequence of events we can do two things: insulate so heavily that no conceivable potential build-up will jump the gap; or reduce the earth resistance to a point where no normal lightning current will produce that high a potential when multiplied times that earth resistance. It is too expensive to do the first thing on 69 kv lines, but higher voltage lines do tend to provide protection against flashovers with their heavy insulation and large spacings.

UNIQUE POWER SYSTEM PROBLEMS—SOLVED

This type of spark-over has been especially troublesome on horizontal post construction, with its limited spacing between phases and down wire. Many cases are being reported of down wires burned in two in the proximity of the phases. This burning is not a result of lightning current, but of power fault current flowing through the spark-over arc from the phase wire into the relatively small down wire. To prevent this, a larger, insulated wire has been used. However, wherever burndowns have occurred, it has been found that the ground resistance reading is very high.

If there is no way to reduce the ground resistance to an acceptable figure, then larger, insulated down wire, and perhaps overinsulation of the phase wires is the best answer. But lowering the ground resistance is preferable.

The second type of problem faced by power engineers is the unshielded line. When lightning hits this line, the "slop" piles up and that sends waves in both directions, but these lines usually have comparatively short span lengths, so it doesn't have to go far to get to a "hole." Since the pole tops are usually the highest points in the vicinity, most of the time lightning strikes right at a pole, so a hole is available right where the lightning hits. If it is a good, big "hole" with a good-sized "pipe" attached with no strainer on the end, almost all of the lightning charge will pass through the hole and very little will be available to ripple away in waves.

But if the strainer is in place and chokes down the flow, it will build up in the pipe until it piles up in the trough. Now this trough is usually on distribution voltage, so the sides of the trough (insulation) are not very high. The slop doesn't have to pile very high before it begins to spill over the sides. And you can imagine what a mess that will make. So what we do is to cut a notch in the side of the trough at a point where we don't mind the slop escaping. We control the place where the slop escapes out of the trough onto the ground. And by the depth of the notch in the side of the trough, we control how high the slop piles up before it escapes through the notch.

The notch is a lightning arrester which spills over at a voltage which is lower than the insulation capability of the line.

The bottom of the arrester is connected to a down wire which is connected to earth at the foot of the pole. If our hole and pipe are large enough and if the strainer is not too constrictive, the slop flows out until the pile-up is exhausted. But if the strainer (earth resistance) is too constrictive, the slop will back up in this pipe and will pile up at the hole and this will be a source of sloshing over. Instead of a solution, we now have a problem. Again we see the importance of getting a good earth ground. Our lightning arrester is helpless without a low-resistance ground under it.

Many times I've had servicemen complain that an arrester "failed," because it exploded. It is more likely that the arrester had a dandy low-resistance ground under it and when Big "L" came along, this good ol' ground soaked up that lightning charge like a drop of water on a dry sponge. The lightning current was just more than a well-designed arrester could take.

We are all fortunate to have available to us many well-designed system components which we can employ in our efforts to keep lightning from damaging equipment. Only one component in our arsenal is completely up to us. That is good grounding. All of the other well-designed components are impotent if not matched up with good, low-resistance grounds.

The first step in obtaining a good ground is to measure the ground resistance with a good Ohmmeter. One that produces an a-c current from a battery source is best.

Understand this one thing. You need to get a measurement of the earth resistance at the base of the pole or tower. Therefore, it is imperative to disconnect the down wire from the ground rod or butt ground. On towers, the shield wire must be disconnected from the down wire and insulated from the tower. If you don't do this, you will be measuring the parallel value of resistance of the earth in the vicinity and all the other earths connected to the neutral, or shield wire, whichever.

Slop waves take time to flow from the point where the slop is being poured into the trough to poles or towers hundreds of feet away. Like the slow-talking gal from South Carolina, before she could say she didn't, she had. Before a lightning surge can get to a remote point to cause damage, it has probably done blown all Hades out of the point it first hit. The reaction of a high resistance-earth ground at the "hit" point will have already been experienced by the line at that point before the first wave gets to the next nearest grounding point. So, it makes little difference what the earth resistance is at any point other than the point you are measuring.

UNIQUE POWER SYSTEM PROBLEMS—SOLVED

It is good to record the Ohmmeter reading above the cut in the down wire at the foot of the pole, then below the cut. You'll find that the reading above the cut may be very low, while that below the cut is very high, indeed. The reading below the cut determines whether lightning current will produce damaging voltage buildup, or not.

Three things you can do to improve your system's reaction to lightning:
1. Improve your grounds;
2. Improve your grounds;
3. Improve your grounds.

24

It was one of those wet, cold March days with a lot of wind and a steady drizzle of freezing rain. It had been sleeting all night and there was a buildup of a half-inch of ice on the wires.

Ice-laden tree limbs had been crashing down over our lines and now, at 8:00 a.m., it began to look like our 34.5 kv transmission system was going to begin to give 'way under its load of ice.

The boss came in and rushed over to me. "Tell me how much load it will take to melt that ice off the 34.5 kv lines in an hour." The idea was to push enough current through the wires to produce enough $I^2 R$ losses to melt the ice.

Out of engineering school only nine months, I was as nervous as a nun in a sporting house. I had twenty minutes to come up with the answer.

I began to search frantically through my conversion tables and found that one Btu raises one pound of water one degree Fahrenheit. Also, the heat of fusion is 144 Btu per pound of ice.

The temperature outside was about 30°F. We had to raise the temperature of the ice 2°F and then melt it.

We didn't have to melt all of the ice; only a path upward from the conductor just as wide as the conductor and ½-inch thick. The conductor was 2/0 ACSR, .447 inches in diameter. .447 × .50, or .2235 square inches was the area of the ice to be melted.

I decided to use 1000 feet as my unit of length. .2235 ×

12" × 1000' = 2,682 cubic inches. That much ice to melt per 1000 feet of wire.

One pound of water = 27.68 cubic inches. The density of ice is .91 times the density of water, so there are 27.68/.91 = 30.4 cubic inches to a pound of ice.

There were 2682 cubic inches of ice to be melted from 1000 feet of wire, so 2682/30.4 = 88 pounds of ice to be melted from 1000 feet of wire.

It takes 2 Btu to raise a pound of ice from 30°F to 32°F and 144 Btu to melt it, so we needed 146 Btu per pound of ice.

My conversion table showed that one kilowatt-hour equals 3413 Btu. To determine the kwh to melt a pound of ice, we divide 146 Btu/lb by 3413 Btu/kwh and get .0428 kwh/lb of ice.

88 pounds of ice per 1000 feet × .0428 kwh per pound of ice equals 3.77 kwh/1000 feet of wire, or 3.77 kw of I^2R losses maintained for an hour.

$$I^2R/1000 = kw \quad I^2 = 1000 \, kw/R.$$

1000 feet of 2/0 ACSR has a resistance of .134 Ohms.

$$I^2 = 1000 \times 3.77/.134 = 28,134$$
$$I = 170 \text{ amperes}$$

On 34.5 kv, we had to switch load to get 170 × 34.5 × $\sqrt{3}$ = 10,160 kva on this line to melt the ice off in one hour.

We did. It did.

In fact, with the help of the wind, short sections of ice cracked loose from the rest and dropped off the line when the melting had reduced the thickness of the ice immediately above the wire to a thin shell. Then the adjacent sections began to slide along the wire toward the center of the span and bump into other sections. This speeded up the shedding process.

It was a real satisfaction to me to see my hastily scribbled figures turn into reality. And the pat on the back from my boss was worth a lot to a nervous young engineer.

From what we have just been through, we can develop an equation for determining quickly how much current will be required in any case to melt the ice off conductors. Let's get off our ice and get at it!

Since the heat of fusion is predominant, and since sleet usually forms at about 30°F, and we can assume we have an hour to get it off, we can start with the information we have already derived, with no changes.

UNIQUE POWER SYSTEM PROBLEMS—SOLVED

Let's start from the end and work back to the beginning, then turn around and go back again.

1. $I^2 R \div 1000 = kw$
2. $I = \sqrt{1000 \ kw/R}$
3. $I = \sqrt{1000 \times kwh \ per \ 1000 \ ft \ of \ wire/R}$
4. kwh per 1000 ft = lbs of ice per 1000 ft × .0428 kwh/lb
5. lbs of ice per 1000 ft = cu inches/30.4
6. cu inches = 12 in./ft × 1000 ft × thickness of the ice × diameter of the wire = 12,000 TD
5. lbs/1000 ft = 12,000 TD/30.4
4. kwh/1000 ft = 395 TD × .0428 = 17 TD
3. $I = \sqrt{1000 \times 17 \ TD/R}$
   $= \sqrt{17{,}000 \ TD/R}$
   $I = 130\sqrt{TD/R}$

   T = Thickness of the ice in inches
   D = Diameter of conductor in inches
   R = Resistance per 1000 ft in Ohms

To assure ourselves that our little formula works, let's put the parameters from my icing problem through it and see if it produces the right answer:

T = .5
D = .447
R = .134
$I = 130\sqrt{TD/R} =$
  $130\sqrt{.5 \times .447/.134} =$
  $130 \times \sqrt{1.67} =$
  $130 \times 1.29 = 168$ amperes. Check.

We usually get worried about icing when it gets about one-half-inch thick, so we can proceed to develop a table for the current required to melt one-half inch of ice off several standard wire sizes. Our formula, with T = .5 constant, now becomes $I = 130 \times .707\sqrt{D/R}$, or $92\sqrt{D/R}$.

ACSR

	D	R	D/R	$\sqrt{D/R}$	I
795 MCM	1.093	.0222	49.2	7.0	644
556 MCM	.927	.0318	29.2	5.5	506
477 MCM	.858	.0372	23.1	4.9	451
397 MCM	.783	.0445	17.6	4.28	394
336 MCM	.721	.0527	13.7	3.78	348
266 MCM	.649	.0664	9.8	3.2	294
4/0	.563	.0835	6.75	2.6	239
3/0	.502	.1050	4.78	2.22	204
2/0	.447	.1330	3.37	1.87	172
1/0	.398	.1675	2.38	1.57	144
1	.355	.2120	1.675	1.32	121
2	.316	.2680	1.18	1.11	102
4	.250	.4240	0.59	.78	72

COPPER

	D	R	D/R	$\sqrt{D/R}$	I
1000 MCM	1.152	.01179	98.0	9.9	911
750 MCM	.997	.01520	65.6	8.1	745
500 MCM	.811	.02219	36.6	6.05	557
300 MCM	.629	.03647	17.3	4.16	383
4/0	.522	.05149	10.1	3.18	293
3/0	.464	.06483	7.2	2.68	247
2/0	.414	.08166	5.1	2.26	208
1/0	.368	.1029	3.58	1.89	174
1	.328	.1297	2.53	1.59	146
2	.320	.1635	1.96	1.40	129
3	.2294	.2062	1.11	1.05	97
4	.2043	.2548	.80	.89	82
6	.162	.4052	.40	.63	58

Note that this ranges from two amperes per MCM for small wire to one ampere per MCM for large wire.

25

Sometimes you can be too neat. For example, we were to hook up a high current circuit made up of two conduits with three 500-MCM conductors in each conduit on a riser pole. Two conductors had to be connected to each phase. The two conduits were side-by-side, so a lineman had connected number-one and number-two wire in the first conduit to phase A, number-one and number-two wire in the second conduit to phase C, and the number-three wires from both conduits to phase B.

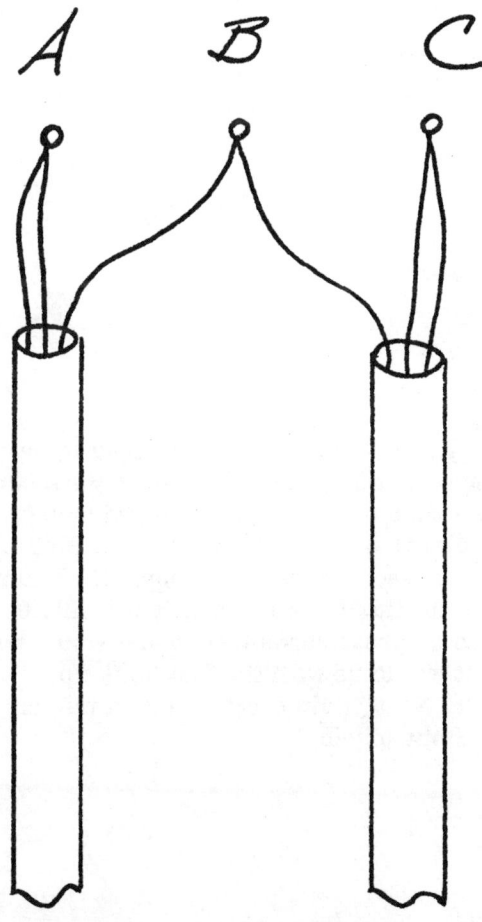

That's about as neat an arrangement as you could dream up. The only trouble is that as soon as you loaded those wires up with current, the steel conduit would heat up and damage the insulation on the wires.

Look at the conduit on the left. It will carry two units of A phase current and one unit of B phase current. At any one instant in time, the flux surrounding these two wires will be produced by the sum of these two currents and it will flow in the wall of the steel conduit, because it would much rather ride in steel than in air, its only other choice. The sum of these two currents would be:

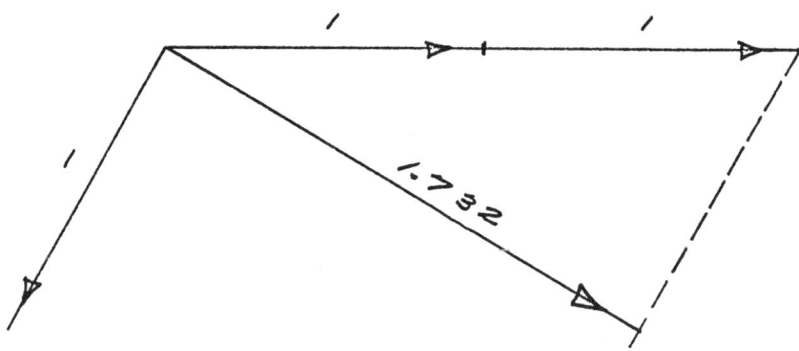

1.732 units of flux would be produced to heat up the conduit wall. The same thing would happen in the second conduit.

The wires should have been connected messy, like the illustration on the next page. Now we have one unit of A, B, and C phase current in each conduit, and since their sum at any one instant is zero, the flux is zero. Actually it would be the result of any unbalance of phase currents, but that would be only a fraction of what it would be with the first hook-up.

And "Mr. Neat" didn't leave enough tail length to reconnect them without splicing!

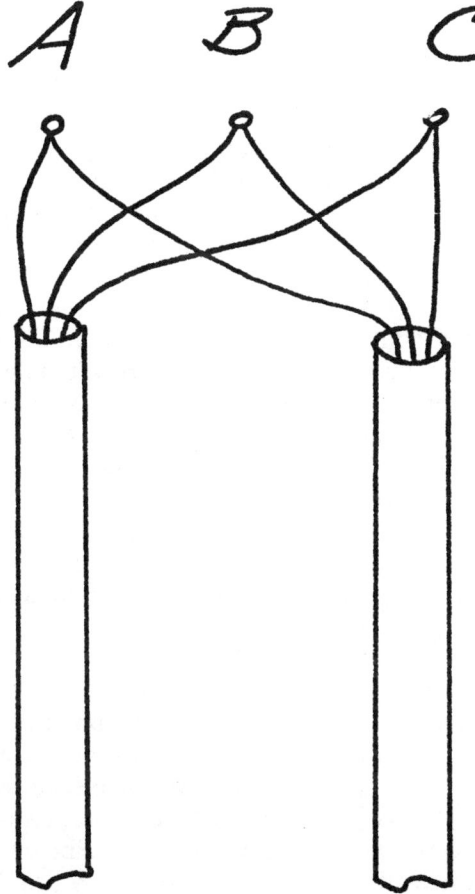

26

A really intelligent high school graduate with thirty or forty years experience as foreman and construction superintendent in the electric utility industry is indispensable to an electrical engineer to provide both sage council and exasperating, cruel, mind-grinding challenge.

These men seem to fall into two basic categories: One, we

have the man who can always come up with a solution to any problem; He's the "can-do" type.

Then, we have the man who, if given sufficient time, can always come up with a problem for any solution. He's the "It can't be done—noway" type. And he's the guy you can count on to keep you out of big trouble. He'll pinpoint the flaw in your plan and cause you some minor embarrassment before you go off at a tangent and cause yourself permanent mortification.

At first glance, it appears that all you need do is get these two characters together and simply act as referee and together the three of you could come up with the solution to any problem. Not so. It would be more like Nero pitching a Christian between a lion and a tiger. These two types just don't mix. You have to treat with each one independently. Go to "Can do." Get a solution. Take it to "It can't be done." Watch him rip it up. Then take all the pieces back to "Can do" and see if he can reassemble it as a usable product. I assume you contribute your part all along the way, of course.

One day ol' Mr. "Can't" called me with what he was sure was the all-time stumper. The perfect crime. No possible solution. He was just drooling with delight. He finally had me. No way out. Ha, ha, ha!

Years ago an old utility manager explained to me that the way he solved any really tough problem was to put it in his desk drawer and leave it there until it solved itself. He was the first man in the outfit to get involuntary early retirement. He was somewhat of a problem, himself. It seems there was another man in our top management who put tough problems in the trash basket.

So, you see, there is no problem that doesn't have some solution.

However, the problem Mr. "Can't" threw at me may have been a first, because I'm not sure I really solved it. But I think I did.

He was having trouble with badly unbalanced voltages on the three phases at the end of a mile-long 266.8 MCM 4-kv feeder circuit whenever a 300-kvar bank of switched capacitors came on the line.

The voltages at the substation were fairly well balanced:

UNIQUE POWER SYSTEM PROBLEMS—SOLVED

$$A - 124$$
$$B - 124$$
$$C - 125\tfrac{1}{2}$$

They remained unchanged whether the capacitor bank was switched off or on.

But the voltages at the end of the line were:

	Caps Off	Caps On
A	120	119
B	119	123
C	123.5	127

The capacitor bank was hooked wye and was grounded. The currents to each phase were normal when the bank was switched on.

The load currents, with the capacitors off, averaged about 160, 160, 120 (A, B, C) for the first half-mile and 80, 80, 50 (A, B, C) for the second half-mile.

This rather severe current unbalance suggested a neutral shift due to the voltage drop in the neutral return circuit. I was assured that there was no open in the neutral anywhere and that all down wires to pole grounds were intact.

I suppose if we accept all of this given data as fact, we must admit that there is no solution to this problem. If there is no neutral (ground return) impedance (to speak of), we can have no appreciable voltage drop in the residual circuit.

However, cynical as it may sound, you just cannot ever accept as gospel the information you are given from the top of a pole by a lineman with bifocals overdue for changing. If it doesn't make sense, you can be sure it ain't so.

In this case, I rode over the circuit myself, and saw numerous joints and splices in the neutral which could possibly be open, or at least be presenting a high resistance to the flow of current. On top of this, the weather had been dry, so the ground return was not at its best.

In the light of all this, I decided to assume that the return impedance was equal to the neutral wire impedance. Of course, it probably varied all along its length, depending on where the bad joints were, and where the ground resistance was bad, or good.

UNIQUE POWER SYSTEM PROBLEMS—SOLVED

Also, the neutral was common to both primary and secondary, so some voltage drop, or rise, was present in some spans due to secondary neutral current. A 12-kv circuit crossed this 4-kv circuit. Some of its neutral current may have hitchhiked a ride on the 4-kv neutral for a way, producing neutral drop which for our purposes appeared as increased return impedance.

Unbalanced phase current will always tend to produce a neutral shift, resulting in high voltage on one phase. Capacitor current tends to accentuate this effect. Let's see what that effect is in this example:

We know what the voltages were at the substation, so we can find the voltages at the end of the line by subtracting the voltage drop in each phase from the voltage at the substation. The voltage drop on each phase is the sum of the voltage drop on the phase wire plus the voltage drop of the return path.

From my knowledge of the load, I assumed a power factor lagging 30°.

The neutral was #4 copper.

The impedance of #4 copper is .70 + j. 40 ohms for half-a-mile. 266.8 MCM is .175 + j .33 ohms per half-a-mile.

$$\#4 - .70 + j.40 = .806\underline{/29.7}$$
$$266 - .175 + j.33 = .374\underline{/62.1°}$$

In the first half-mile, with capacitors off:

$$I_a = 160\underline{/-30°} = 138.6 - j\,80.0$$
$$I_b = 160\underline{/-150°} = -138.6 - j\,80.0$$
$$I_c = 120\underline{/+90°} = 0 + j\,120.00$$
$$I_n = 0 - j\,40.0 = 40.0\underline{/-90°}$$

In the second half-mile, with capacitors off:

$$I_a = 80\underline{/-30°} = 69.3 - j\,40.0$$
$$I_b = 80\underline{/-150°} = -69.3 - j\,40.00$$
$$I_c = 50/+90° = 0 + j\,50.00$$
$$I_n = 0 - j\,30.0 = 30.0\underline{/-90°}$$

The voltage drop in the neutral is:

$$40.0\underline{/-90°} \times .806\underline{/29.8°} + 30.0\underline{/-90°} \times .806\underline{/29.8°}$$
$$= 32.24\underline{/-60.2°} + 24.8\underline{/-60.2°} = 56.42\underline{/-60.2°}$$

UNIQUE POWER SYSTEM PROBLEMS—SOLVED 109

The voltage drop in "A" wire is:
  160/−30° × .374/27.9° + 80/−30° × .374/27.9°
  = 59.84/−2.1° + 29.92/−2.1° = 89.76/−2.1°

The voltage drop in "B" wire is:
  160/−150° × .374/27.9° + 80/−150° × .374/27.9°
  = 59.84/−122.1° + 29.92/−122.1° = 89.76/−122.1°

The voltage drop in "C" wire is:
  120/+90° × .374/27.9° + 50/+90° × .374/27.9°
  = 44.88/117.9° + 18.7/117.9° = 63.58/117.9°

"A" wire drop	89.76/−2.1°	=	89.7 − j 3.3
NT drop	56.42/−60.2°	=	28.0 − j 49.0
			117.7 − j 52.3
"B" wire drop	89.76/−122.1°	=	−47.7 − j 76.0
NT drop	56.42/−60.2°	=	28.0 − j 49.0
			−19.7 − j 125.0
"C" wire drop	63.58/117.9°	=	−29.8 + j 56.2
NT drop	56.42/−60.2°	=	28.0 − j 49.0
			− 1.8 + j 7.2

The voltages at the substation were:
  A − 124 × 20 = 2480/0° = 2480 + j 0
  B − 124 × 20 = 2480/−120° = −1240 − j 2150
  C − 125.5 × 20 = 2510/+120° = −1255 + j 2175

"A" minus V.D.$_a$
  2480.0 + j 0
  − 117.7 + j 52.3
  2362.3 + j 52.3 = 2363.0/1.3°

"B" minus V.D.$_b$
  −1240.0 − j 2150.0
  +   19.8 + j 125.0
  −1220.2 − j 2025.0 = 2365.0/238.9°

"C" minus V.D.$_c$
  −1255.0 + j 2175.0
  +   1.8 − j   7.2
  −1253.2 + j 2167.8 = 2504.0/+120.0°

On a 120-volt base:
A = 118.2
B = 118.2
C = 125.2

Now let's switch those capacitors on and see what effect they have:

First half-mile:

I_a $\qquad\qquad$ 138.6 − j 80.0
$I_{a,\,cap}$ 42/+90° = $\quad$ 0 + j 42.0
$\qquad\qquad\qquad\qquad$ 138.6 − j 38.0 = 144.0/−15.3°

I_b $\qquad\qquad$ −138.5 − j 80.0
$I_{b,\,cap}$ 42/−30° = $\quad$ 36.4 − j 21.0
$\qquad\qquad\qquad\qquad$ −102.1 − j 101.0 = 143.0/224.7°

I_c $\qquad\qquad$ 0 + j 120.0
$I_{c,\,cap}$ 42/−150° = − 36.4 − j 21.0
$\qquad\qquad\qquad\qquad$ − 36.4 + j 99.0 = 105.5/110.2°

Second half-mile:

I_a $\qquad\qquad$ 69.3 − j 40.0
$I_{a,\,cap}$ 42/90° = $\quad$ 0 + j 42.0
$\qquad\qquad\qquad\qquad$ 69.3 + j 2.0 = 69.3/1.7°

I_b $\qquad\qquad$ − 69.3 − j 40.0
$I_{b,\,cap}$ 42/−30° = $\quad$ 36.4 − j 21.0
$\qquad\qquad\qquad\qquad$ − 32.9 − j 61.0 = 69.3/241.7

I_c $\qquad\qquad$ 0 + j 50.0
$I_{c,\,cap}$ 42/−150° = − 36.4 − j 21.0
$\qquad\qquad\qquad\qquad$ − 36.4 + j 29.0 = 46.5/+141.5°

The neutral current in the first half is:

I_a $\quad$ 138.6 − j 38.0
I_b $\quad$ −102.1 − j 101.0
I_c $\quad$ − 36.4 + j 99.0
I_n $\qquad$ 0 − j 40.0 = 40/−90°

The neutral current in the second half is:

UNIQUE POWER SYSTEM PROBLEMS—SOLVED

I_a 69.3 + j 2.0
I_b −32.9 − j 61.0
I_c −36.4 + j 29.0
I_n 0 − j 30.0 = 30/−90°

The voltage drop in the neutral is:

40/−90° × .806/29.8° + 30/−90° × .806/29.8° =
32.24/−60.2° + 24.18/−60.2° = 56.42/−60.2° =
28.0 − j 50.0

Note that this is the same as with no capacitors, since caps produce essentially no neutral current.

The voltage drop on "A" wire is:

144.0/−15.3° × .374/27.9° + 69.3/1.7° × .374/27.9° =
53.86/+ 12.6° + 25.92/+ 29.6° =

$$\begin{array}{r} 52.6 + j\ 11.7 \\ 22.6 + j\ 12.8 \\ \hline 75.2 + j\ 24.5 \\ \text{Nt. drop} \quad 28.0 - j\ 50.0 \\ \hline 103.2 - j\ 25.5 = \text{V.D.}_a \end{array}$$

"A" minus V.D.$_a$

$$\begin{array}{r} 2480.0 +\ 0.0 \\ -130.2 + j\ 25.5 \\ \hline 2349.8 + j\ 25.5 = 2350.0/+ 0.6° \end{array}$$

The voltage drop on "B" wire is:

143.0/224.7 × .374/27.9° + 69.3/241.7° × .374/27.9° =
53.48/107.4 + 25.92/−90.4° =

$$\begin{array}{r} -16.0 + j\ 51.0 \\ 0 - j\ 26.0 \\ \hline -16.0 + j\ 25.0 \\ \text{Nt. drop} \quad 28.0 - j\ 50.0 \\ \hline 12.0 - j\ 25.0 = \text{V.D.}_b \end{array}$$

"B" minus V.D.$_b$

$$\begin{array}{r} -1240.0 - j\ 2150.0 \\ -\ \ 12.0 + j\ \ \ 23.9 \\ \hline -1252.0 - j\ 2126.1 = 2467.3/−120.5° \end{array}$$

112 UNIQUE POWER SYSTEM PROBLEMS—SOLVED

The voltage drop on "C" wire is:

105.5/+110.2 × .374/27.9° + 46.5/141.5° × .374/27.9° =
39.46/138.3° + 17.4/+169.4° =

$$\begin{array}{r} -29.5 + j\ 26.2 \\ \underline{-17.1 + j\ \ 3.2} \\ -46.6 + j\ 29.4 \end{array}$$

Nt. drop $\underline{\ \ 28.0 - j\ 50.0\ \ }$
$-18.6 - j\ 20.6 = $ V.D.$_C$

"C" minus V.D.$_C$

$$\begin{array}{r} -1255.0 + j\ 2175.0 \\ \underline{+\ \ \ 18.6 + j\ \ \ \ 20.6} \\ -1236.4 + j\ 2195.6 = 2520.0/+119.4° \end{array}$$

On a 120-volt base:

A = 117.5
B = 123.4
C = 126

Let's look at both answers together:

	No Caps	Caps On
A	118.2	117.5
B	118.2	123.4
C	125.2	126.0

The difference between the highest and lowest voltage with no capacitors is 7.0 volts. With the capacitors switched on the difference becomes 8.5 volts, a 21% increase.

The answer to this problem is *not* to turn the capacitors off, although that's just what we did in this case until we could provide a permanent solution. The answer is to balance the load on the phases and go to work on that neutral and grounds. The problem ceased as soon as we did just that.

27

He was confused, bewildered and a little hurt. He'd been supervising contractor line-crew work for years. Prior to that he'd worked as a lineman and a serviceman. He'd phased out primaries and secondaries many times and now, here he was, trying to phase together the primaries from two substations, both fed from the same transmission system and they just wouldn't go.

"Let's see, now," I said. "Primaries from sub number one feeding the double dead-end from the east and primaries from sub number two from the west, right?"

"Right. And I had a lineman using a hotstick voltmeter measuring voltages across the double dead-end. He read 0, 7500 volts and 15,000 volts."

"Are you sure he read it right?"

"Yes, I'm sure!"

At this point I made a mistake. As I've told you, it is the cardinal rule in solving any electrical riddle to assume that the data you are given are inaccurate. But this was my best man, I just couldn't bring myself to doubt him.

So I spent some hours plotting vector diagrams assuming all sorts of crazy substation hookups to try to validate his voltage readings. No luck. It just couldn't be.

"What was the weather when you took those readings?" I asked.

"It was drizzling rain."

"You know, those hotstick voltage meters are tricky, especially in wet weather. Try it again as soon as the weather dries up and I think you'll get different readings."

He did and he did.

His new readings were:

$A_1 - A_2$ 7.5 kv
$A_1 - B_2$ 7.5 kv
$A_1 - C_2$ 15   kv

This was more like it. Either the lineman misread the first time, or he didn't make a good connection.

"You'll have to go to both subs and see which one you can reverse two 69-kv phases on. Neither sub is connected with any other sub, so it can be either one. However, most of our three-phase substation transformers are fed ACB, CBA, or BAC, so try to do it that way, if you can."

He checked and found that sub number one was ABC and sub number two was ACB. He reversed sub number one and the two systems went together just fine.

Of course, you'll quickly notice that by reversing two leads on the high side of the substation we reversed the rotation on the 12-kv system fed out of that sub, so we had to reverse two leads on the low side to keep the three-phase motors from running backwards.

Now, if we reversed once on the high side and once on the low side, why didn't we have the same difficulty in phasing out that we did before?

Or, better still, why couldn't we have just reversed two leads on the low side only, to correct the situation? Why mess with the 69-kv side at all?

And why do I say that a three-phase transformer presents a more difficult problem than three single-phase transformers?

First, let's make sure we understand fully the problem of rotation.

If we connect a three-phase motor to a source, it will rotate one of two ways—right, or wrong. Some motors don't care which way they turn because they are hooked to loads which operate equally well whether turning clockwise or counterclockwise. However, most three-phase motors are hooked to loads which must turn in a particular direction. For instance, a fan must turn in the proper direction or it will suck instead of blow.

If the motor rotates in the wrong direction when we connect a source to it, all we have to do is reverse two of the phase wires and reconnect. Then it will rotate in the other, correct, direction.

Or we could go back to the primary phases and reverse two of them. This would work just as well for the motor in question, but it would reverse rotation of all the other motors connected to that primary system past that point. We sure don't want to do that.

We could go back to the transmission system and reverse two phases. This will correct the rotation of our motor in question, but it will cause all the other motors past that point on the transmission system to run backwards.

We could go back to the generator and reverse two leads and cause some real, widespread problems.

We could run the generator backwards, spinning it in reverse rotation. Now, we see that the rotation of the system is set by the rotation of the generator. If it turns clockwise, all three-phase motors fed by it turn clockwise if the phases are carried straight through all the way from the generator to the motors. But we can reverse two phase wires anywhere along the way and cause all the motors past that point of reversal to rotate in the opposite direction from all the motors ahead of that point.

Rotation is either right, or wrong. No matter how the phases come into the motor, if it spins in the right direction, the rotation is correct. If it spins in the wrong direction, the rotation is wrong and we must reverse two phase wires, somewhere, to correct the rotation.

Now, let's look at some standard substation transformer hookups and determine the effects of bringing in the high-side phases in different ways. Since almost all distribution systems now are fed from Delta-Wye hookups, we'll use Delta-Wye. Let's assume 69-kv to 12-kv.

There are only six possible ways the 69-kv can come into the high side of the transformation:

If A is the left phase, then the other two can come in BC, or CB;

If B is the left phase, then the other two can come in CA, or AC;

If C is the left phase, then the other two can come in AB, or BA;

ABC, ACB, BCA, BAC, CAB, or CBA.

From observation we can see that three of these produce motor rotation in one direction and three in the other direction.

Group I is ABC, BCA, CAB;
Group II is ACB, CBA, BAC.

This simply means that a 69,000-volt motor with its terminals 1, 2, 3, hooked to a Group I system either 1-2-3, A-B-C; 1-2-3, B-C-A; or 1-2-3, C-A-B; will spin in the opposite direction from the same motor hooked to a Group II system, either 1-2-3, A-C-B; 1-2-3, C-B-A; or 1-2-3, B-A-C.

It also means that a 10-horsepower 240-volt three-phase

motor fed from a Group I substation will reverse rotation if the primary circuit feeding it is switched off of a Group I sub and reconnected to a Group II sub.

This can be corrected by reversing two leads into the motor itself, but this won't help all those other three-phase motors on the system. If this is the only three-phase bank on the system we might reverse two primary leads into the bank.

But it is likely that the system feeds numerous three-phase banks, so we want to fix our problem by making a switch at the substation so that one switch will correct rotation of all the three-phase motors on the system at once. This we can do by swapping two phase leads between the transformer bank and the main bus.

We've got our motors all spinning in the right direction, and we can switch our primaries back and forth from one sub to the other without reversing motor rotation. But we cannot parallel the two subs. We have to deenergize our primaries each time before switching to the other sub.

In order to be able to switch a section of primary circuit from one sub to another, without first deenergizing it, we must arrange to get both subs into one group or the other.

Now, let's go back through and make sure we fully understand this business of motor rotation. It's easy to get motor rotation and rotation of phase vectors confused. Since the two are related, it makes it very easy to get wound up in your underwear trying to keep it all straight in your head.

First, look at the generator. Essentially it has three coils physically spaced 120 degrees apart with a lead brought out from each coil. Three leads; let's mark them 1-2-3. Let's crank the generator slowly by hand until a voltmeter connected to lead number 1 indicates maximum positive voltage. We know that at that point, we will find that voltmeters hooked to leads 2 and 3 will read negative 50% voltage.

Now, crank the generator in the direction it will be running when actually in service and watch the voltmeters on lead 2 and lead 3 to see which one reaches maximum positive voltage first.

Let's arbitrarily christen lead 1, phase A. If lead 2 reached maximum positive voltage before lead 3, then lead 2 is phase B, and lead 3 is phase C. If lead 3 reaches maximum positive

UNIQUE POWER SYSTEM PROBLEMS—SOLVED

voltage before lead 2, then lead 3 is phase B and lead 2 is phase C.

Now, let's connect a motor directly to the generator leads. This motor is identical to the generator in design so that its 1-2-3 leads physically match the 1-2-3 leads of the generator. Obviously, it will rotate in the same direction as the generator if 1 is hooked to 1, 2 to 2, and 3 to 3. Also, rotation will be the same if 1 is connected to 2, 2 to 3, and 3 to 1. Or, if 1 is connected to 3, 2 to 1, and 3 to 2.

However, rotation will be in the reverse direction if 1 is connected to 1, 2 to 3, and 3 to 2; or 2 to 2, 3 to 1 and 1 to 3; or 3 to 3, 1 to 2, and 2 to 1.

Let's look at it another way: If lead 1 is at maximum positive and lead 2 is on the rise, the rotor will be pulled-pushed in one direction; but if lead 3 is on the rise, the rotor will be pulled-pushed in the opposite direction.

A motor is affected similarly. If we hook our supply phase wires to leads 1-2-3 in such a manner that when lead 1 is at maximum positive, lead 2 is rising in voltage, the motor will rotate in a direction opposite to its rotation if the supply phase wires are so connected that lead 3 is the "riser" when lead 1 is at the maximum positive.

It doesn't make any difference where on the system supplied by this generator we connect our motor, 12-kv, 4-kv, 480 volts, 240 volts, or whatever, the same logic prevails. Connect A-B-C to either 1-2-3, 2-3-1, or 3-1-2, and it will run in one direction. Connect A-B-C to 1-3-2, 3-2-1, or 2-1-3, and it will rotate in the opposite direction.

If all transformations on the system were Wye-Wye, this would be duck soup. A would be A from generator to motor. B would be B, and C would be C.

But transform Delta-Delta, Delta-Wye, or Wye-Delta, and we drag our pivot foot. We lose our bearings and get hopelessly confused. Add to this some of the conflicting transformer connection diagrams supplied in textbooks and in manufacturers' publications and we are ready to throw up our hands in frustration. Sometimes, just throw up—period.

Let's look at the various transformer connections and voltage vector diagrams and develop them logically in the light of our previous discussion.

But first, why get all excited about this subject? Well, because we must face two problems which directly affect the operation of our system;

1. *Motor rotation:*—Motors must rotate in the right direction to do the work they are designed for.
2. *Vector relationship:*—In order for two systems to be connected together, whether transmission, primary, or secondary, their vector diagrams must be identical.

Before proceeding, it will be helpful if you understand this. When we go through a Wye-Wye transformation, or through an "American Standard" Delta-Delta transformation the vector relationships remain unchanged. A is still A, B is B, and C is C. A-B is still A-B, B-C is still B-C, and C-A, still C-A. Not so on any other transformation. We find it useful to arbitrarily name the phase wires on other transformations, A-B-C, but it ain't necessarily so. Our low side phases are brand new phases, not always vectorially identical with the high side phases. We name low side "a" to be the phase wire which is most nearly like phase A on the high side, or which we think should logically be called "a." Even when "a" is actually identical with high side phase A, phase a-b is not identical to high side phase A-B. It is a brand new phase which we haven't seen before.

Now, let's start with voltage vectors. What is the relationship between the sine wave and the vectors?

Take a bicycle wheel and attach a baseball to the rim. Then attach another at a point 120° from the first and a third 120° from the other two. Hold it up in front of a white wall, with the axle parallel to the wall. Spin the wheel. Shine a floodlight on the wall so that the wheel and baseballs make a shadow on the wall. Now walk from left to right holding the spinning wheel and balls in front of you. The balls' shadows will trace three sine waves on the wall equivalent to phases A, B, and C. Stretch a tape from the hub of the wheel to each ball. The tapes represent phases A-B-C. The balls are the arrows at the end of each vector.

You see that both the sine wave shadow and the wheel represent the phase action. We are getting two views; one from the side, one from the front.

We arbitrarily place the arrow on the ball end of the vector.

The arrow is on the end of the N-phase vector that rotates. Once chosen this way, we must stick with this representation from now on. Otherwise, you have a Russian novel.

Let's assign a vector to each coil of the generator, the coils connected in Wye. (They could be connected Delta and it would make no difference. There would still be an N-phase voltage force.)

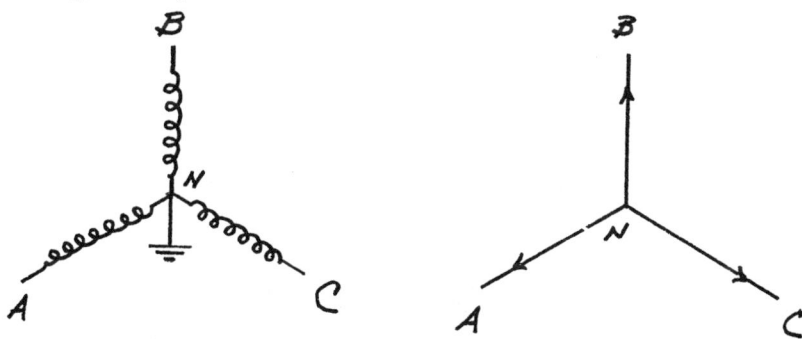

The vector representing phase A-B is the sum of A-N plus N-B. A-N is the opposite of N-A, so we must reverse this vector.

B-C is B-N plus N-C.
C-A is C-N plus N-A.

The complete picture, then:

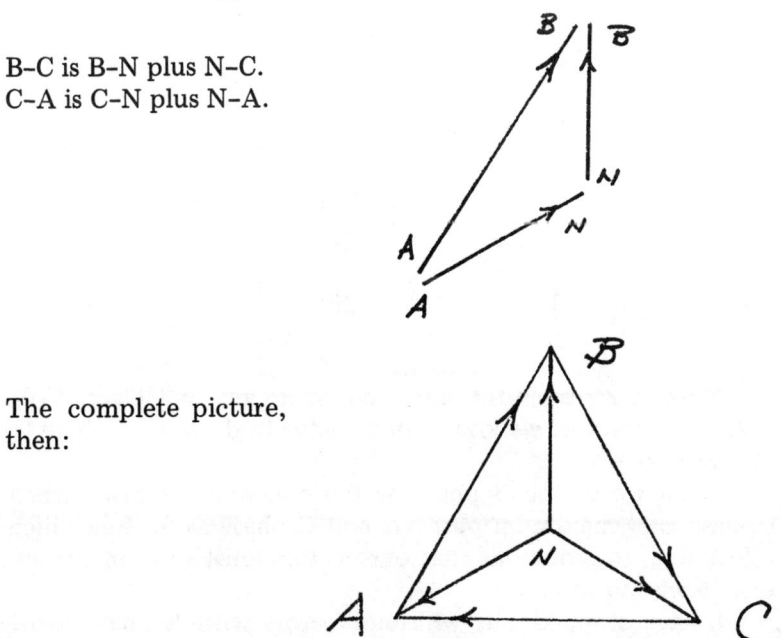

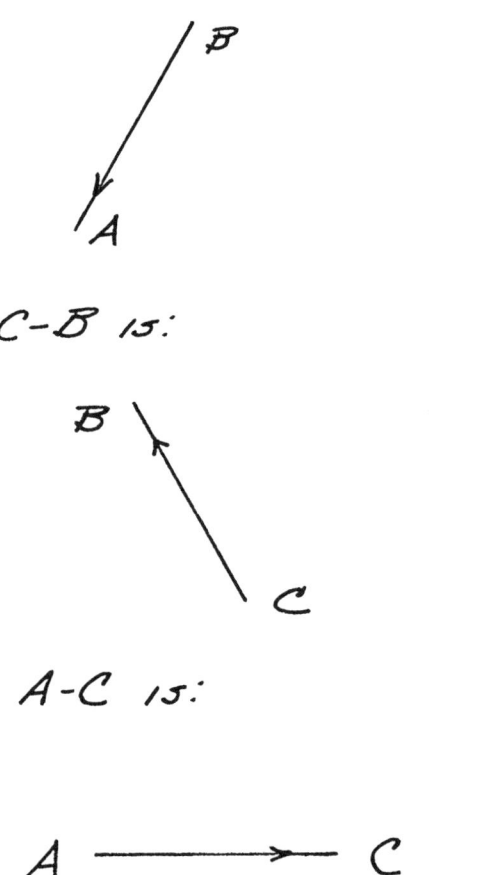

Now, if we trace through a coil from any one letter, A-B-C-N, to any other, we know what vector to draw, and where to attach the arrow.

Since we define B phase as the one which is rising when A phase is at maximum positive, and C phase as the one which is declining, it is obvious that our vectors must whirl in a counterclockwise motion.

I have stopped our whirling vectors with B phase in its

UNQIUE POWER SYSTEM PROBLEMS—SOLVED 121

maximum position. We can stop them anywhere, as long as we stay with that position throughout the entire investigation. Since our phases always retain the same relationship to each other, our manipulations will always produce identical results no matter in what position the vectors are "frozen." Most publications on transformer connections use the B-up position, so I will do the same here so that you can more easily compare our work to theirs.

At this point, I emphasize that the rotation of our phase vectors must not be confused with motor rotation. The three phase vectors represent the voltage which is impressed on the three wires we name A-B-C. We arbitrarily name one wire "A" at the generator terminals. Then "B" automatically becomes the wire whose voltage is rising when "A" is at maximum positive, and "C" is the other one. Eventually these three phases arrive at a motor. We connect A phase wire to one of its three terminals; say, terminal number one. We can then connect B phase wire to terminal two or to terminal three. (C takes whatever is left.) If B is connected to terminal two, the motor spins in one direction. If it is connected to terminal three, it spins in the opposite direction.

Often I've heard it said that the rotation of a particular transformer bus is "positive," or "negative." This is a misconception that causes hours of grief for many a young, or old, electrical engineer. Transformers do not rotate. They do not "have rotation." They are static devices. Only motors rotate. Transformers produce rotation of motors. Once the transformer connection is chosen, all the motors on that system must be connected to conform with that connection. As each new motor is attached to the system, it is observed and if its rotation is correct, O.K. If not, we must reverse two of its leads to reverse its rotation.

We are now ready to leave the generator terminals and take our voltage vectors out over the system, finally arriving at a three-phase motor connected to secondaries located many miles away.

The voltage of the generator is transformed to transmission level by a Wye-Wye connection, probably an autotransformer, so our vector diagram remains intact. Let's say our transmission voltage is 69 kv. Our vector diagram is:

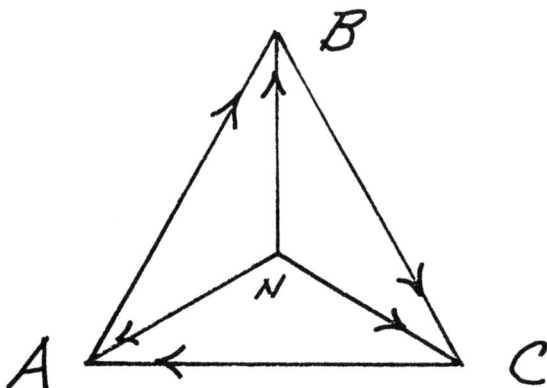

We know we don't carry a neutral on our transmission lines, but the neutral voltage point is there every foot of the way.

Now, let's apply the voltage to a 69/12 kv transformation and see what comes out on the 12 kv side. We'll use a three-phase Delta-Wye transformer with subtractive polarity.

Polarity. If you tie the adjacent high and low side terminals of a transformer together and apply a voltage across one of the coils, do we read on a voltmeter connected between the other high and low voltage terminals more, or less, voltage than was applied to the coil? If more, we say the transformer is additive. If less, subtractive.

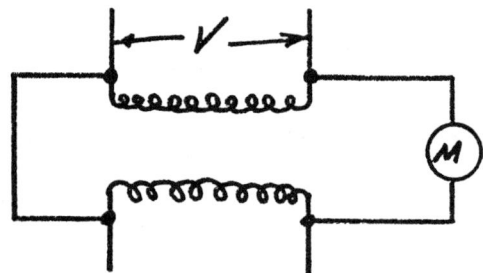

In terms of voltage vectors, what significance does polarity have?

UNIQUE POWER SYSTEM PROBLEMS—SOLVED 123

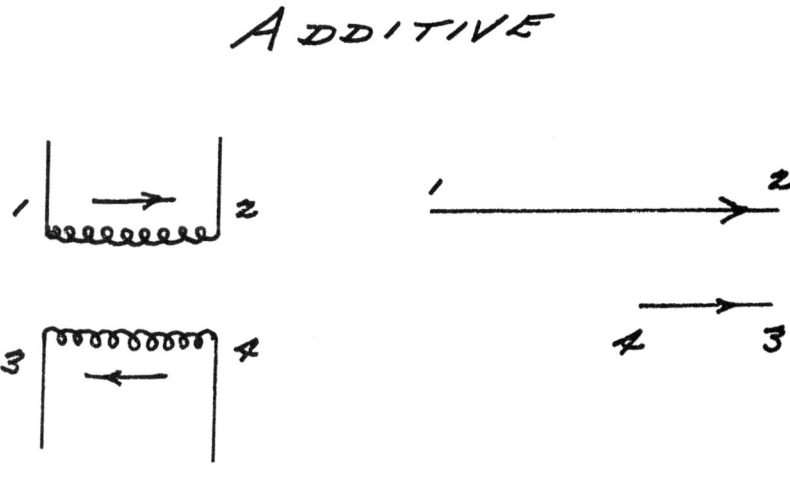

The arrows on the transformer diagram tell us that if, at any one instant, the induced voltage is pushing from 1 to 2, induced voltage will, at the same instant, be pushing from 4 to 3. The voltage push at that instant is from 1 to 2 and from 4 to 3. We connect 1 to 3 and measure additive voltage from 2 to 4. This tells us that 1-2 and 4-3 voltages are pushing in the same direction so their vectors must be in the same direction, as shown. Note that if we tie point 1 on the first vector to point 3 on the second vector, we get a new vector which is the sum of the two original vectors. Note that the load current in the primary coil will flow in the opposite direction from the induced voltage. It is pushed on by the *impressed* voltage. As the secondary load current increases, it produces flux in the core which opposes the induced primary voltage just enough so that the difference between the impressed and induced primary voltages pushes current in the primary equaling the secondary current, divided by the turns ratio.

124 UNIQUE POWER SYSTEM PROBLEMS—SOLVED

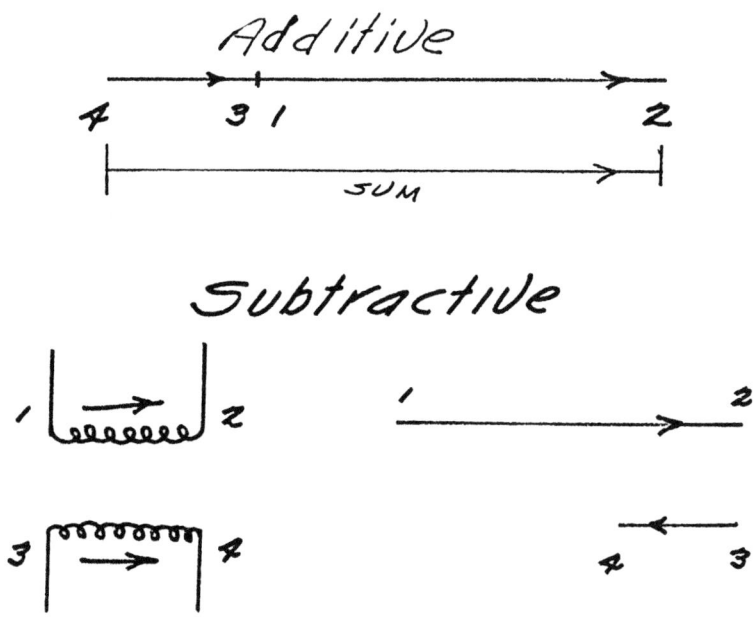

At the same instant that induced voltage is pushing from 1 to 2, it is also pushing from 3 to 4. We connect 1 to 3 and measure subtractive voltage from 2 to 4.

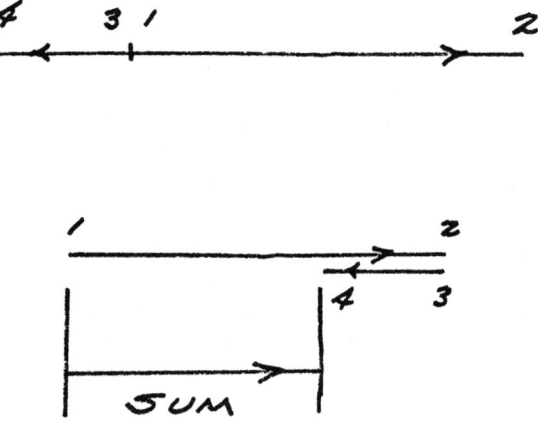

The standard diagram for a three-phase Delta-Wye transformer is:

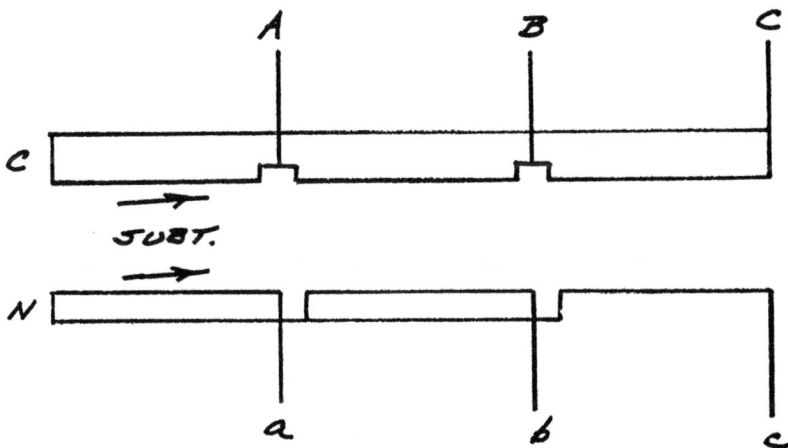

Remember, the voltage applied is:

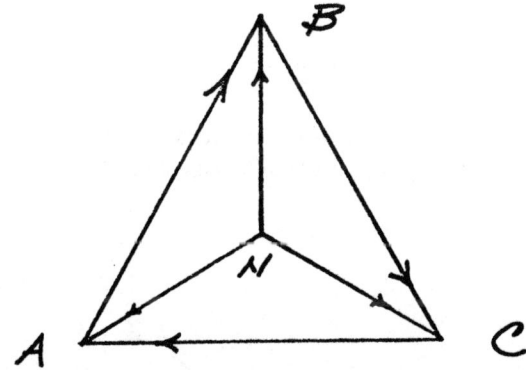

The voltage from C to A is identical to the voltage from N to a, ignoring transformation ratio (which we shall do throughout). So, the vector for N to A is:

A-B is like N-b, so:

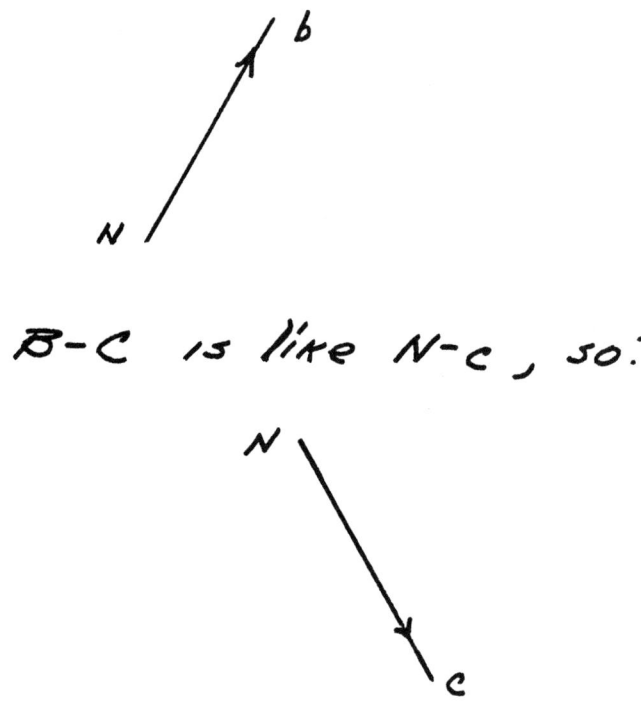

N is common to all three vectors, so we can form the complete diagram by tying all the N's together.

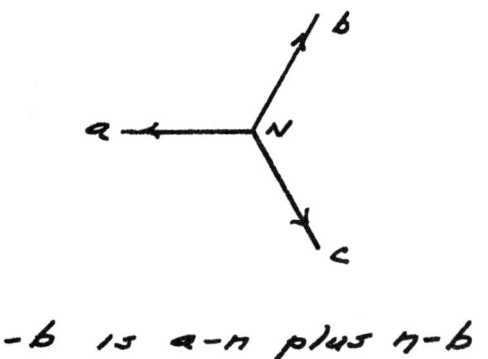

UNIQUE POWER SYSTEM PROBLEMS—SOLVED

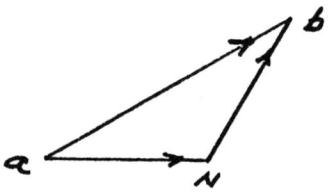

b-c is b-N plus N-c

c-a is c-N plus N-a

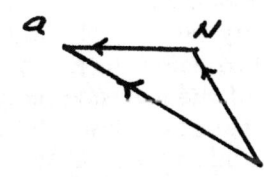

ASSEMBLING:

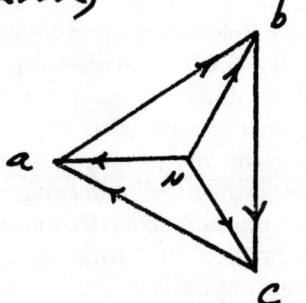

Let's Compare with our High Side Diagram:

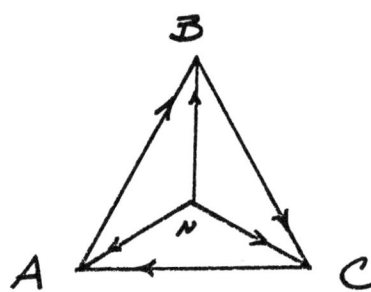

We see that our secondary phase vectors have rotated minus 30°. A better way to say this is to admit that our old A-B-C system is gone forever. We have an entirely new system of voltages and voltage vectors. We now must go through the same christening process we did at the generator terminals. We have three brand new phases, 1-2-3. Which one shall we call "A"?

Since the low side, left-hand phase is produced by high side coil A-C, it is logical *not* to call that phase "b," since "B" was not involved. Why call it "a"? Why not "c"? Well, because it would be illogical to name the arrow end of the low side vector for the tail end of the high side vector. It makes more sense to name the arrow end of the low side vector for the arrow end of the parent, high side vector. Thus, we call the right-hand, low side terminal of the left transformer, "a." The middle transformer becomes "b," and the right transformer, "c." The low side vector diagram is labeled correspondingly.

Actually, once we get "a" labeled on the vector diagram, "b" and "c" are determined, because the vectors must rotate counterclockwise in accordance with our original assumption. If we stand at the bottom of the page and mentally rotate the vectors until "a" passes, then "b" must pass next, and then "c".

Our low side "a" phase reached its maximum positive value 30° after our high side "A;" "b" follows "B" 30° and "c" follows "C" 30°.

UNIQUE POWER SYSTEM PROBLEMS—SOLVED

We brought our 69 kv phases into the high side in a Group I configuration. We use A-B-C. What happens if we use the other two Group I configurations?

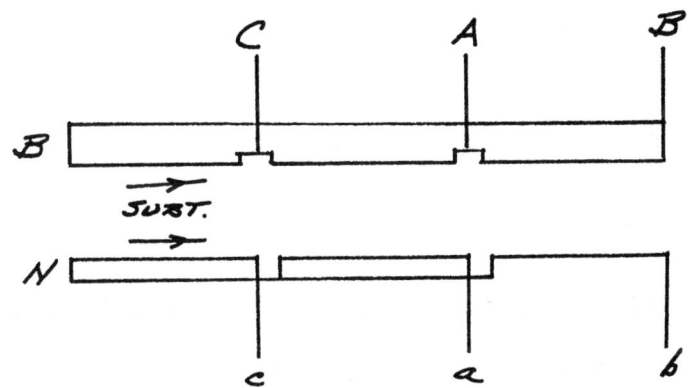

THE VOLTAGE APPLIED IS STILL:

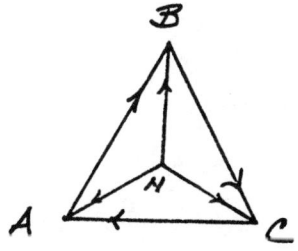

B-C IS N-c. C-A IS N-a. A-B IS N-b.

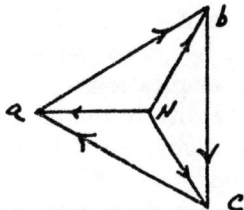

WE GET AN IDENTICAL DIAGRAM TO THE ONE WE GOT WITH A-B-C.

130 UNIQUE POWER SYSTEM PROBLEMS—SOLVED

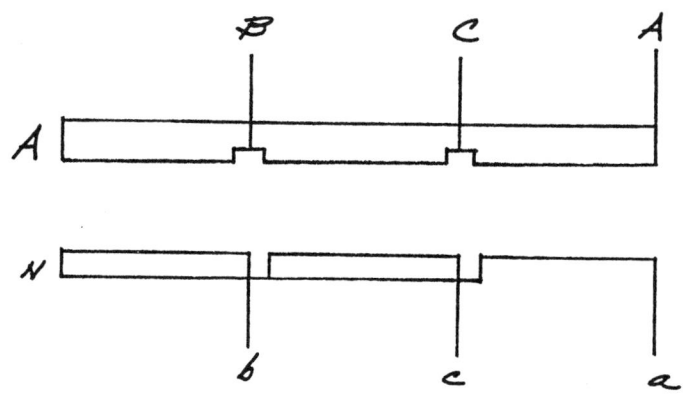

A-B IS N-b. B-C IS N-c. C-A IS N-a.
AGAIN WE GET:

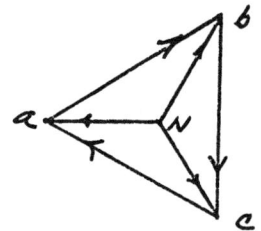

The diagrams are identical for all Group I configurations. But the phase leads to the substation bus have different phase markings. First, we had a-b-c, then c-a-b, then b-c-a. What effect does this have on motor rotation?

Let's look at a motor with terminals 1-2-3 tied respectively to a-b-c in our first configuration. The motor rotation is correct, of course, or somebody would already have reversed two leads. Anyway, let's assume correct rotation. Now, kill the sub and reconnect the high side C-A-B. This means that 1-2-3 is now tied respectively to c-a-b. In the first example, terminal 2 was fed by a phase which was 120° behind the phase feeding terminal 1, and so forth. In the second example, the same thing is true. So rotation remains unchanged.

Again, we see that any Group I produces the same motor rotation as any other Group I.

Also, we see that any Group I substation can be paralled with another Group I substation. Of course, they must be phased out first.

Let's inspect a Group II:

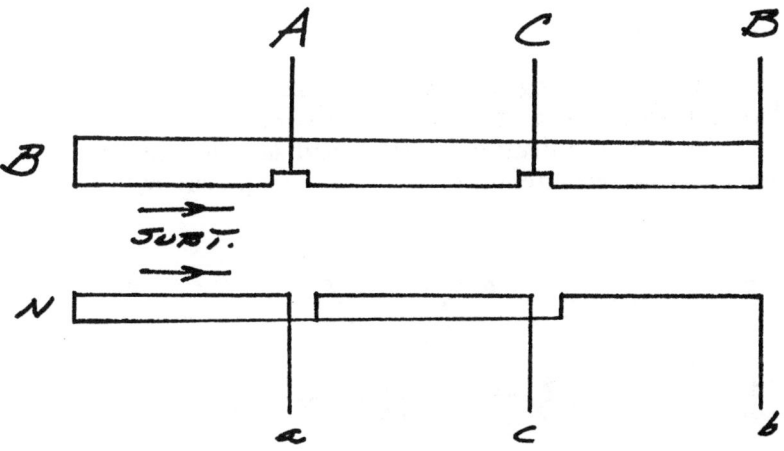

The voltage diagram of the supply transmission system is still:

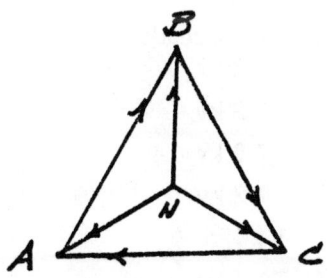

B-A is N-a. A-C is N-c. C-B is N-b.

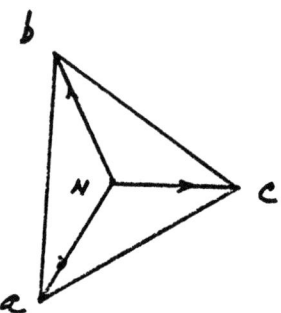

Instead of being rotated —30° from the high side, as with Group I, we see that Group II is rotated +30°.

Let's tie the neutrals together and see what difficulties we run into when we attempt to parallel a Group I sub with a Group II sub:

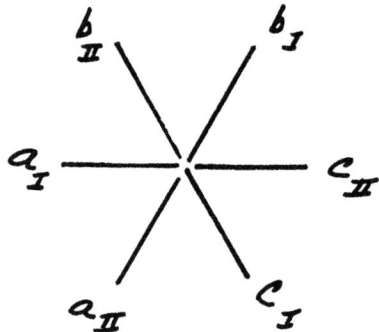

On our 12-kv system we read:

$a_I - a_{II}$	7.2 kv
$b_I - b_{II}$	7.2 kv
$c_I - c_{II}$	7.2 kv
$a_I - b_{II}$	7.2 kv
$a_I - c_{II}$	14.4 kv
$b_I - a_{II}$	14.4 kv
$b_I - c_{II}$	7.2 kv
$c_I - a_{II}$	7.2 kv

$c - b_{II}$ 14.4 kv

Now, we can see vectorially the problem which faced my contractor inspector 'way back at the start of this story. These are precisely the voltage readings he got with his hotstick voltmeter. No wonder he was puzzled!

Remember that I told him to go to one of the two substations and roll two of the incoming 69 kv phases, changing it from a Group I to a Group II, or vice versa. By doing this he produced the same voltage diagram on the low side of both substations. You will also remember that this swap alone would have caused all the motors on that substation to run backwards, so he also reversed two leads between the transformer and the bus. Then he phased out at the double-dead-end where the station 12 kv primaries met, and all was just fine.

We were dealing with three-phase transformers. A three-phase transformer has six coils, three high side coils and three low side coils. That gives us twelve coil ends and twelve leads. But only six leads are brought out of the tank where we can get at them to manipulate and reconnect them. We are limited by this physical property to the solution we actually applied in this case.

However, if you have a sub with three single-phase transformers in it, you can do some tricks.

Let's parallel a 12-kv system, fed from a Group I, three-phase transformer sub, with a 12-kv system fed from a Group II, three single-phase transformer sub.

Our 12-kv vector diagrams are:

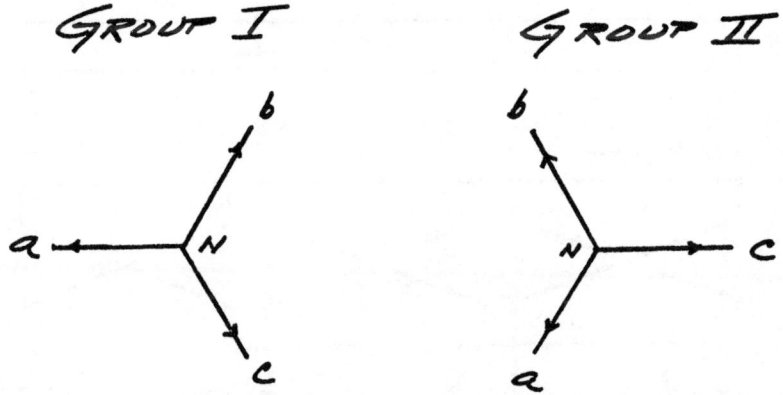

Our Group II transformer connection is:

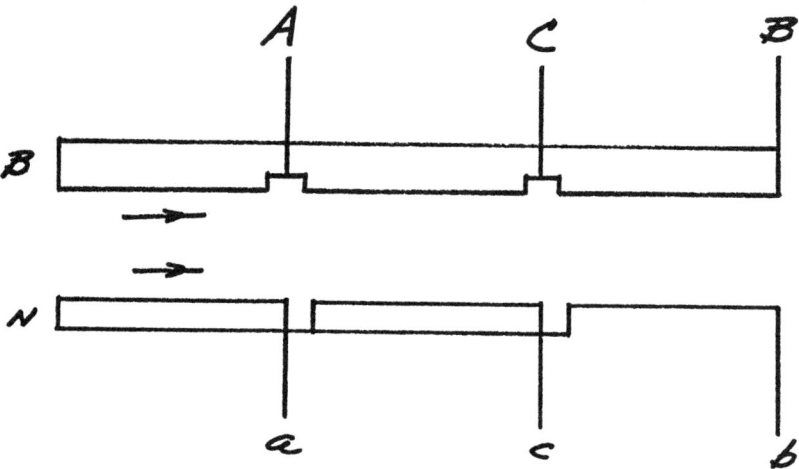

What can we do to this transformer hookup to produce a Group I vector diagram without having to reverse any transmission connections, since this is usually very difficult to do?

Let's try taking the neutral connection apart and connect those coil ends to the phase wires. The coil ends which were previously connected to phase wires we'll now connect together as the neutral:

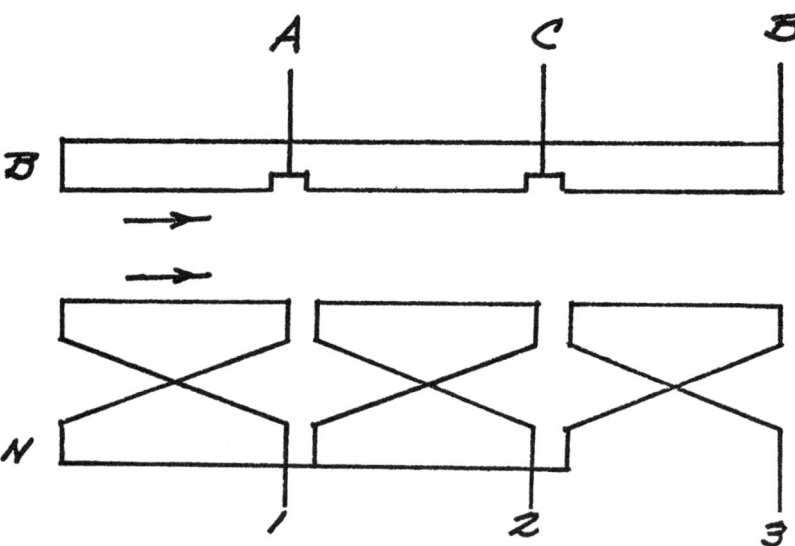

UNIQUE POWER SYSTEM PROBLEMS—SOLVED

Now let's do our phase tracing thing:

B-A produces minus N-a from terminal N to terminal 1. A-C produces minus N-c from N to 2. C-B produces minus N-b from N to 3. And we get the following diagram:

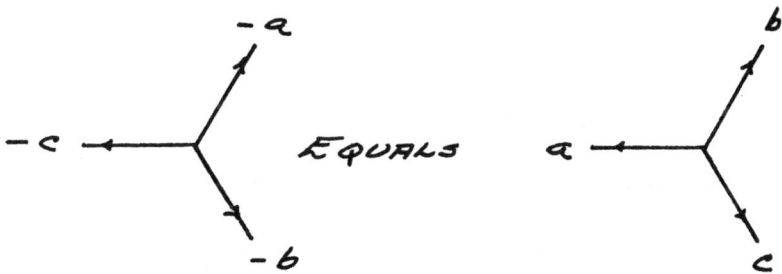

This diagram falls exactly on our Group I diagram. Since terminal 1 is (—) a, and (—) a is coincident with b—so what— we'll just call terminal 1, b. Terminal 2 was (—) c. That is coincident with a, so call terminal 2, a. Terminal 3 was (—) b. That's coincident with c, so call terminal 3, c. We have followed a logical pattern to christen the three terminals and our diagram is now:

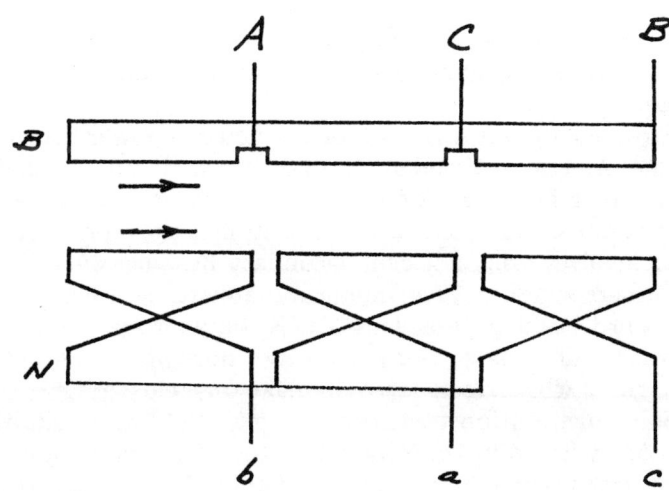

Our vector diagram is now Group I, so we know that the two subs can be paralleled. And you see, we will still have the same motor rotation since we have not changed our phase sequence. We just rotated our vector diagram clockwise 60°, or counterclockwise 300°, if you wish to be difficult about this thing.

If the point on the distribution system at which you intended to put these two systems together, say at a switch, or a double-dead-end, had correct phasing before you reconnected the Group II sub, it won't have now. If before, A from Group I sub, was directly across the dead-end from A (from Group II sub), and B across from B and C across from C, you will now have to go back to a vertical corner, or dead-end, and do some phase-rolling, because Group II sub's phases A-B-C have now become phases B-C-A. But that will probably be easier than trying to twist two incoming 69-kv phases.

Why can't you just reconnect the phases at the sub and avoid any phase-rolling out on the line? Well, you can, if that's easier, because it will not change motor rotation.

We have solved our original problem of parallelling two substations, fed by dissimilar high side phase groupings. This solution can also be applied to paralleling distribution transformer banks.

To complete our study of the effects of various transformer connections, let's walk through each of the possible connections carefully so that you won't be left behind the way I have been left behind by so many explanations I've waded through.

First, let's point out that for distribution power banks, we have the American Standard connection and the Alternative connection. Since most of these banks are now mounted on cluster mounts, the Alternative connection is the more practical since it involves tying adjacent secondary bushings of adjoining transformers together. The American Standard ties each bushing to the third bushing over. This is O.K. on a two-pole platform bank where all leads go straight up to a horizontal bus, but I'd hate to think what a mess it would make on a cluster.

Since distribution transformers up to 167 kva are additive, we'll stick with additive. Remember, though, if you connect a 250-kva transformer in a bank with two 167-kva transformers, you must reverse the leads on the subtractive 250-kva unit to make it, effectively, an additive unit.

Let's begin with the easy one, the Wye-Wye:

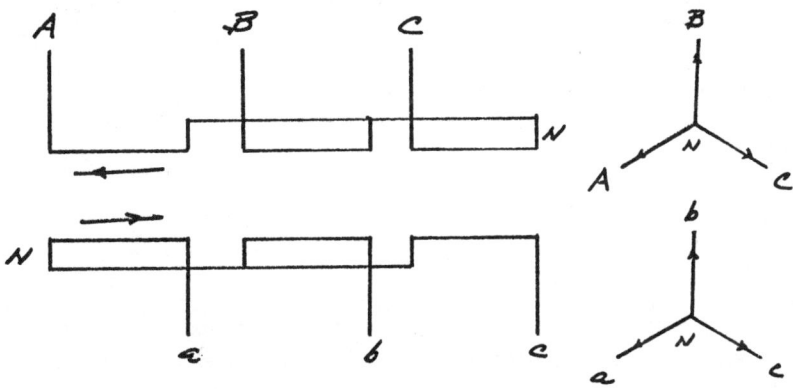

N-A is equal to n-a, N-B equals n-b, and N-C equals n-c.

We can foul this up by bringing our high side leads into the right-hand side, or by taking our low side leads out the left-hand side. Like this:

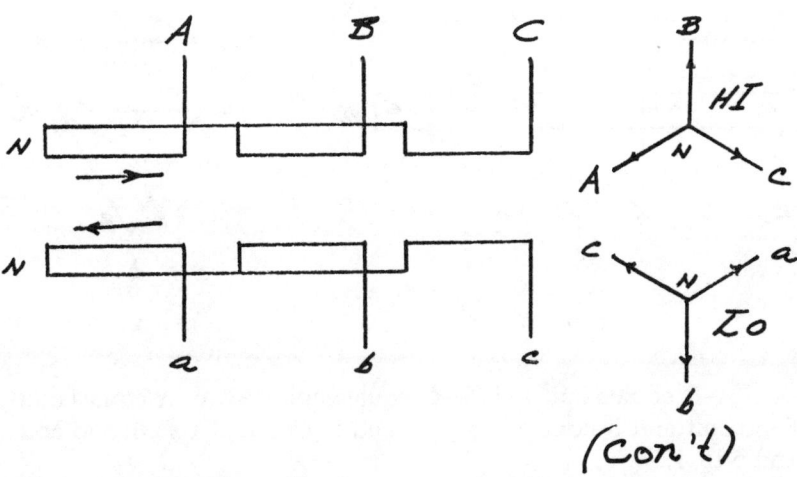

(con't)

138   UNIQUE POWER SYSTEM PROBLEMS—SOLVED

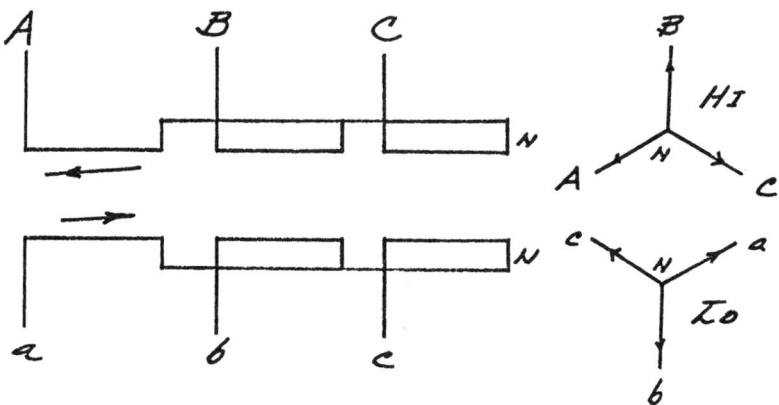

Of course, if our transformers were subtractive, this would present no problem. Our HI and LO voltage diagrams would be identical.

Now, look at Delta-Delta, Alternate connection:

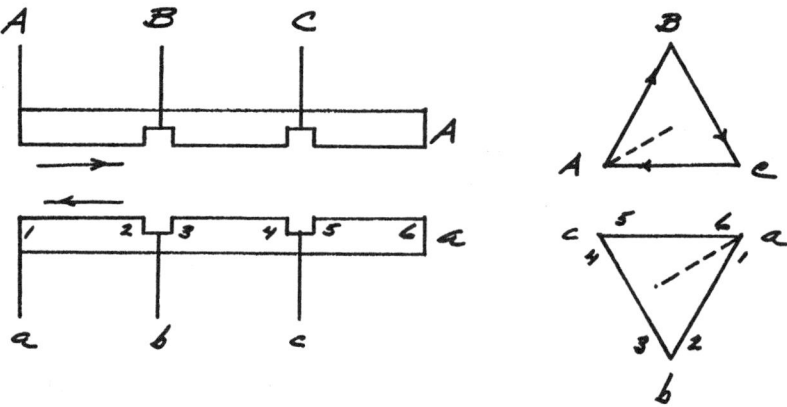

A–B equals coil 2–1. B–C equals coil 4–3. C–A equals coil 6–5. Coil end 3 connects to coil end 2. Coil end 1 to 6, and coil end 5 to 4.

Why do we call point 6-1, a? Well, end 1 is related to high side phase B. Tracing from A to B on the high side coil should be equivalent to tracing from 2 to 1 on the low side, so 2-1 might be called a-b. By similar reasoning, 4-3 might be called b-c, and 6-5, c-a. But point 6 (c) is tied to point 1 (b). Should junction 6-1 be called c or b? There is no more reason to call it one than the other. The obvious answer is to call it a to avoid hurting the feelings of either c or b! Likewise, point 5-4 (a-b), we'll christen c, and point 3-2 (c-a), we'll call b. The rotation is correct. Our new n-a is 180° out of phase with N-A.

Now, look at Delta-Delta, American Standard connection:

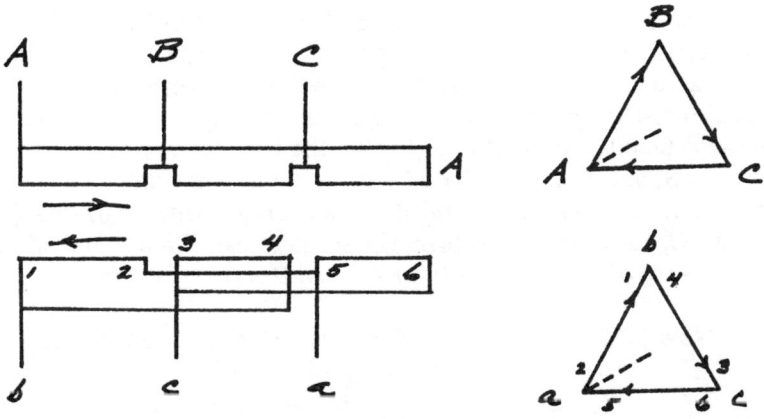

A-B is 2-1. B-C is 4-3. C-A is 6-5. This time we find that 2 and 5, which are tied together, are both a, 1 and 4 are both b, and 3 and 6 are both c, so christening the low side connection diagram and vector diagram is as easy as falling in love.

The rotation is correct and there is no phase shift from high side to low.

Now, look at Wye-Delta, Alternative connection:

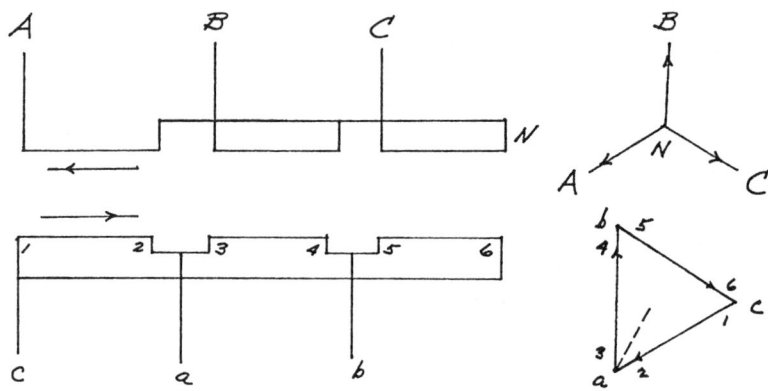

N-A is 1-2. N-B is 3-4. N-C is 5-6. Draw 1-2 just like N-A. Draw 3-4 just like N-B with 3 starting at point 2, since 3 is connected to 2 in the transformer diagram. Draw 5-6 just like N-C, with 5 starting at point 4, since 5 is connected to 4 in the transformer diagram. This brings 6 out at point 1, as it should, because 6 and 1 are tied together on the transformer diagram.

1, 3, and 5 are associated with N, so it is not logical to give them phase names. 2, 4, and 6 are associated with high side A, B, and C, respectively, so it makes sense to call them a, b, and c.

N-a is 30° out of phase with N-A. Rotation is correct.

Now look at Wye-Delta, American Standard connection.

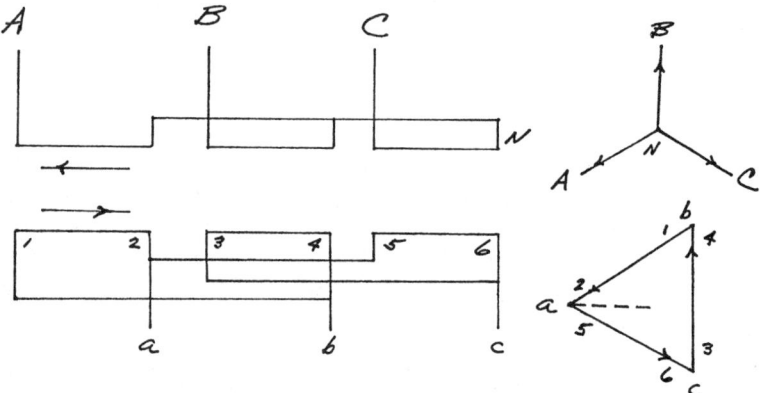

N-A is 1-2. N-B is 3-4. N-C is 5-6. 1, 3 and 5 are associated with N, so we don't find it logical to give them phase names. 2, 4, and 6 relate to A, B, and C, so we call them a, b, and c. The rotation is correct and n-a is 30° out of phase with N-A.

Now let's look at Delta-Wye:

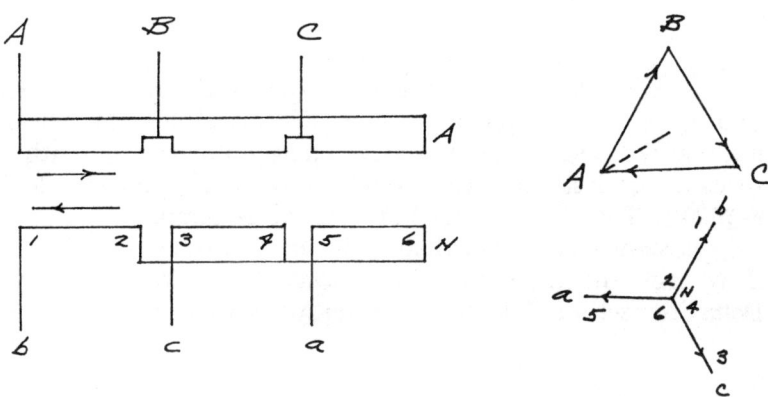

A-B is n-b (2-1). B-C is n-c (4-3). C-A is n-a (6-5). Why call point 1, b? Well, the arrow on the A-B vector is on the B end of the vector. The arrow on the 2-1 vector is on the 1 end of the vector. So it seems logical to call it b.

The rotation is correct and n-a is 30° out of phase with N-A, lagging.

Let's look at the last connection and see what happens if we bring the phase leads out the other side of the secondary coils. Hook 1, 3, 5 together as the neutral and bring 2, 4, 6 out as phases.

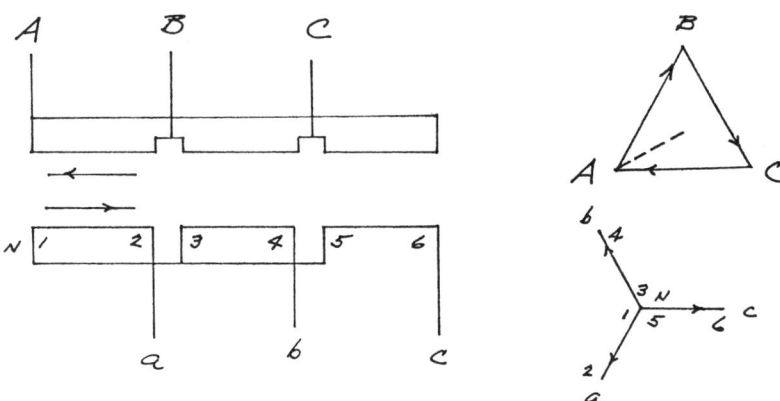

B-A is n-a (1-2). C-B is n-b (3-4). A-C is n-c (5-6). The rotation is correct, but now n-a is 30° out of phase with N-A, leading. We have shifted the vector diagram ahead 60° with respect to our last connection where we used the left-hand leads as phases. These two connections cannot be paralleled.

It stands to reason that we'll run into the same 60° shift if we bring the phases into the right-hand high side of our Wye-Delta connection. Let's try it on the Alternative connection.

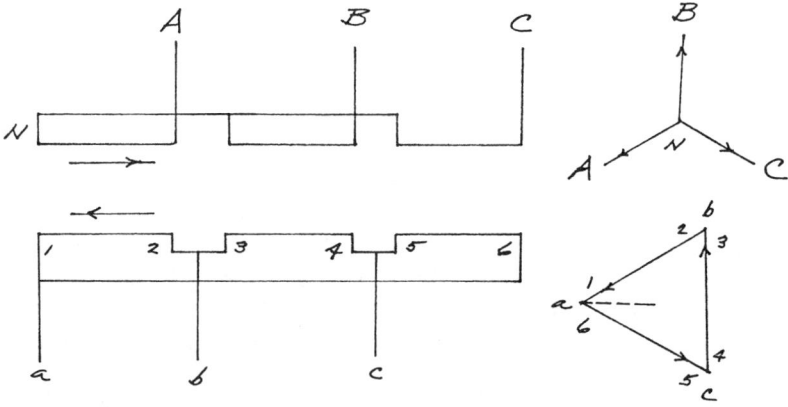

N-A is 2-1. N-B is 4-3. N-C is 6-5. Call 1, a, since the arrow is on the 1 end of the vector. 3 is b and 5 is c.

The original vector diagram was: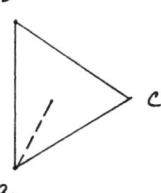

Bringing the high side phase leads in from the right has shifted our diagram 60°, as we suspected, but this time it is 60° behind (clockwise) instead of ahead.

Open Deltas and open Wye-open Deltas can be investigated by simply drawing the closed bank connections and removing one transformer.

Let's summarize what we've discussed in this section.

Phase sequence is determined at the generator. After arbitrarily christening one terminal "A," by convention "B" is the terminal whose voltage is rising at the instant "A" reaches maximum positive. "C" is the terminal whose voltage is going more negative at that instant.

Rotation of motors is determined by whether the phases are connected to the terminals 1, 2, 3 in a Group I (ABC, BCA, CAB) arrangement or a Group II (ACB, CBA, BAC) arrangement.

A Group I primary or secondary can not be paralled with a Group II primary or secondary.

The phases are transformed through a Wye-Wye bank unchanged in the Standard connection.

The phases are transformed unchanged through the American Standard Delta-Delta connection, but are shifted 180° through the Alternative connection.

The Alternative Wye-Delta shifts the phases forward 30° and the American Standard shifts them 30° back.

Delta-Wye shifts the phases 30° forward if the secondary leads are brought out the right, and 30° back if they are brought out the left.

We waded slowly through the construction of voltage diagrams and I am sure you can now do your own thing without any more help from me.

28

Once in a while you're called upon to engineer a job in a way that would make you ashamed of yourself under normal circumstances. For instance, would you recommend that a 50 hp, 240-volt, three-phase motor be supplied from an open Delta bank over five miles of two-phase 4.16-kv line? And what if the line were composed of one #1 copper phase, one #6 copper phase, and one #6 copper neutral? You would recommend adding a third phase, right?

No, you wouldn't. Not in the example I have in mind. You'd just sit down and calculate whether you could make the thing operate successfully on the existing two-phase circuit with transformers as small as possible.

Sounds pretty dumb, right?

Not so dumb when you realize that this motor is to be used as a floodwall pump to be operated for a few days, no more than twice a year. It is used to protect a small rural town from flooding. Your franchise in this town has only two years to run. The town has very little money. They can't afford a reduced voltage starter. The two-phase line is no longer loaded. The main load was located some miles past the pump location and is now served from another source. The line now supplies only scattered farm loads, so there is no justification, other than the proposed pump, for adding a third phase. You don't want to have to go before the somewhat hostile city council and propose that they contribute several thousand dollars to offset your large investment to supply full three-phase to this highly unprofitable load.

So you sit down and calculate the voltage dip which you can expect on this line when that 50 hp pump starts—across the line.

You could calculate it on a full three-phase basis and adjust your answer to take into account the missing phase, but your answer would be worthless because this is anything but a balanced system. We can see immediately that there will be more voltage drop on one phase than on the other. We'll just have to calculate the drop in each phase and in the neutral, independently.

First, we must determine how much current the motor will suck on starting. Then we must determine how much current

UNIQUE POWER SYSTEM PROBLEMS—SOLVED 145

will appear on each primary phase and on the neutral. Then, we can calculate the voltage drop on each phase and on the neutral and with that we can get the total voltage dip on each phase.

A motor such as this will draw, on starting, about six times full load current, if full voltage is maintained at its terminals. From this assumption, we should be able to determine the approximate impedance which this motor will present to our "two-faced" system at the moment we close the switch to attempt to start it. We know we'll have far less than full voltage then, but we will still be facing the impedance we calculate on this basis. The impedance won't change, that's why we call it "impedance" instead of calling it "helpfulness."

We also know that the impedance angle on start will be about 75°.

The full load current is: (Assume power factor is .746 so horsepower equals kva)

$$\frac{50 \text{ kva}}{.240 \times \sqrt{3}} = 120 \text{ amperes}$$

$$120 \times 6 = 720 \text{ amperes } \underline{/-75°}$$

Let's assume Delta connected motor coils. Coil current is line current divided by $\sqrt{3}$.

$$\frac{720 \underline{/-75°}}{\sqrt{3}} = 416 \text{ amperes } \underline{/-75°}$$

Let's draw a sketch of this motor fed from a closed Delta bank, showing currents:

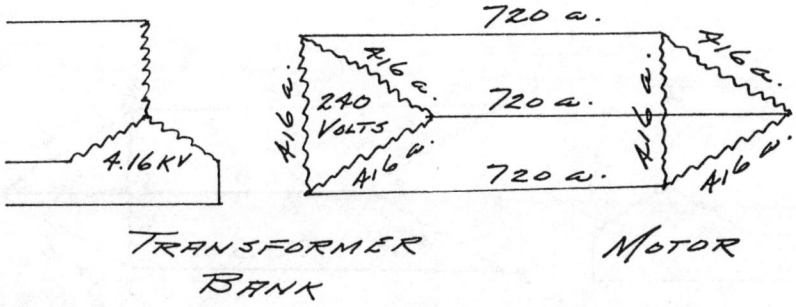

146   UNIQUE POWER SYSTEM PROBLEMS—SOLVED

Each 240-volt transformer coil supplies 416 amperes to its corresponding motor coil. The line currents are the vector sum of two coil currents.

Now, if we remove one transformer coil, the remaining two coils continue to supply 416 amperes to their corresponding motor coils.

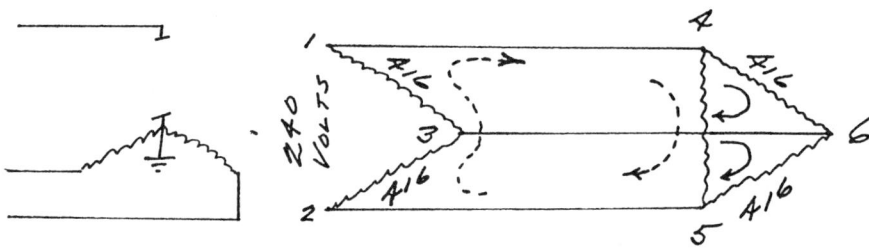

3-1 supplies 4-6, and 2-3 supplies 6-5. 240 volts is still maintained across 1-2. Therefore, full voltage is still maintained across motor coil 4-5, so it takes its current of 416 amperes, too. This 416 amps is still supplied from 1-2, but now there is no coil there, so this current has to pass through coils 2-3 and 3-1 in series, adding vectorially to the 416 amps in 2-3, which is at one phase angle, and to the 416 amps in 3-1, which is at a different phase angle. Now our sketch looks like this:

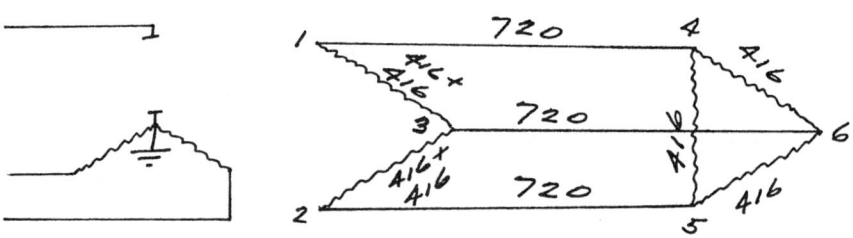

UNIQUE POWER SYSTEM PROBLEMS—SOLVED

The current in each motor coil lags the voltage across that coil by 75°. So, our vector diagram at the motor looks like this:

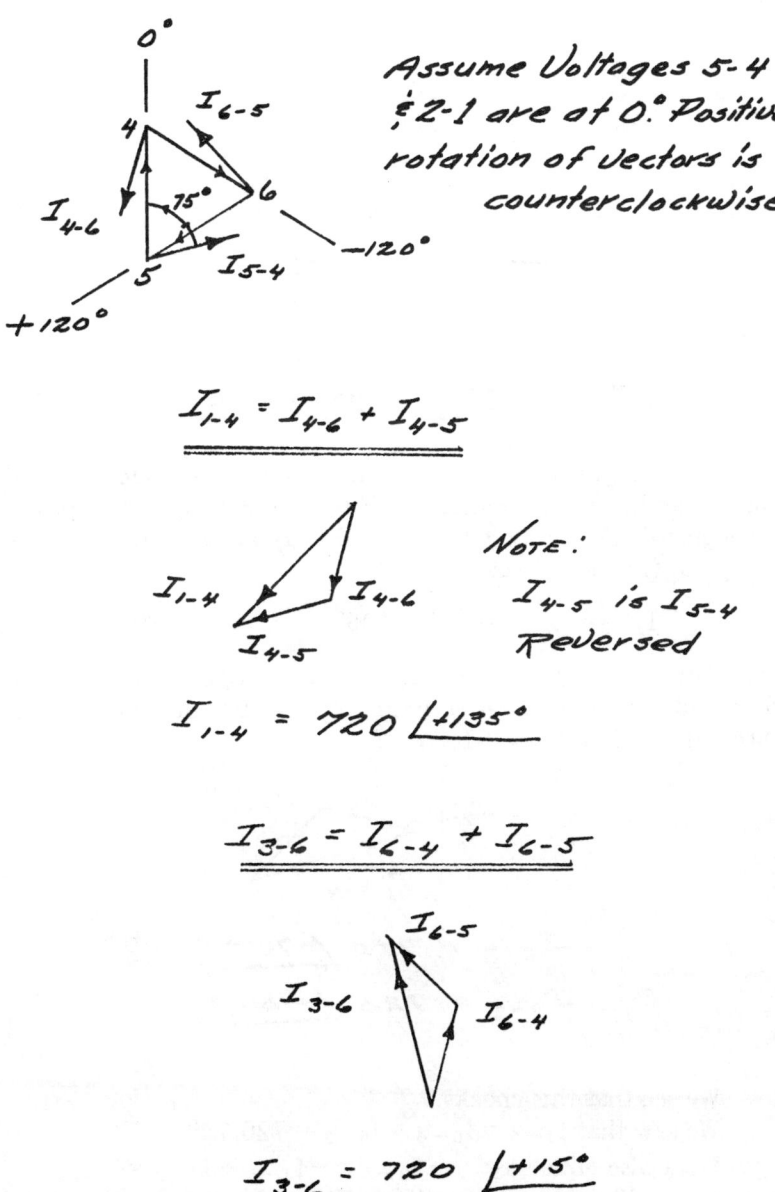

Assume Voltages 5-4 & 2-1 are at 0°. Positive rotation of vectors is counterclockwise.

$$I_{1-4} = I_{4-6} + I_{4-5}$$

NOTE: I_{4-5} is I_{5-4} Reversed

$$I_{1-4} = 720 \angle +135°$$

$$I_{3-6} = I_{6-4} + I_{6-5}$$

$$I_{3-6} = 720 \angle +15°$$

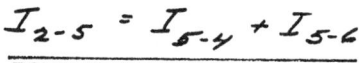

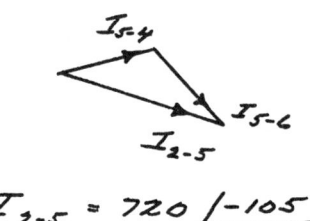

$$I_{2-5} = 720\,\underline{/-105°}$$

$$I_{3-1} = I_{1-4} = 720\,\underline{/+135°}$$

But I_{3-1} also equals $I_{4-6} + I_{4-5}$ since it supplies coil 4-6 current and serves as a series path for the current which passes through coil 4-5. $I_{4-6} + I_{4-5}$ has already been calculated to be $720\underline{/+135°}$, which checks.

$$I_{3-2} = I_{2-5} = 720\underline{/-105°}$$

But I_{2-3} equals $I_{6-5} + I_{4-5}$ since it supplies coil 6-5 current and serves as a series path for the current which passes through coil 4-5. $I_{6-5} + I_{4-5}$ is:

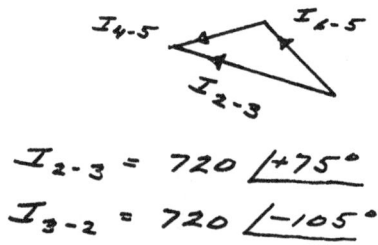

$$I_{2-3} = 720\,\underline{/+75°}$$
$$I_{3-2} = 720\,\underline{/-105°}$$

We see that this checks.
We saw that $I_{3-6} = I_{6-4} + I_{6-5} = 720\underline{/15°}$
I_{3-6} also equals $I_{1-3} + I_{2-3} = -I_{3-1} + I_{2-3} =$
$\quad 720\underline{/-45°} + 720\underline{/+75°} = 720\underline{/+15°}$

UNIQUE POWER SYSTEM PROBLEMS—SOLVED

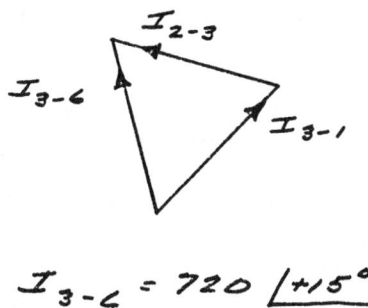

$$I_{3-6} = 720 \,/\!+\!15°$$

And this checks.

So we have determined the transformer and motor coil currents which would be experienced if the motor were started across-the-line with full voltage maintained; and just for fun, we have examined what changes would occur to these currents if we lop one transformer out of the circuit and establish an Open Wye–Open Delta transformation.

We found that nothing changes except the currents in the two remaining transformer coils. They jump in value from 416 to 720 and shift in phase angle.

Note that these secondary coil currents would be transformed to the high side and would add vectorially to form the neutral current.

The impedance of each motor coil is equal to the voltage divided by the current, which lags that voltage in each coil by 75°.

$$\frac{240 \text{ volts}}{416/\!-\!75°} = .577/\!+\!75° \text{ ohms}$$

Now our diagram looks like this:

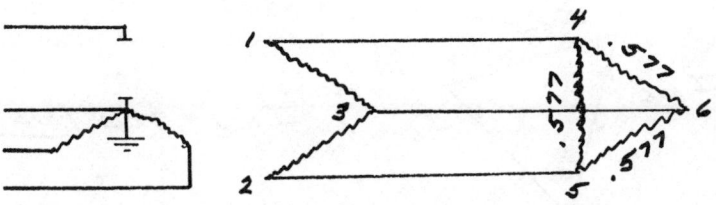

150 UNIQUE POWER SYSTEM PROBLEMS—SOLVED

How on earth can we convert this three-phase impedance into two single-phase impedances to be operated upon by our two-phased system voltage? Why, let's just replace them with two impedances from 4-6 and from 6-5 which produce the same currents, at the same phase angles, which we have previously calculated. Remove the impedance from 4-5 altogether so it will not produce any further confusion and difficulty!

If we do this, $I_{4-6} = I_{1-4}$ and $I_{2-5} = I_{5-6}$ and $I_{3-6} = I_{6-4} + I_{6-5}$. As soon as we determine I_{4-6}, we can find the new equivalent impedance of coil 4-6. Finding the current in 6-5 allows us to determine the new equivalent impedance in that coil. The new equivalent impedance of coil 4-5 is infinity.

$$I_{4-6} = I_{1-4} = 720\underline{/135°}$$

$$Z_{4-6} = \frac{V_{4-6}}{I_{4-6}} = \frac{240\underline{/-120°}}{720\underline{/+135°}} = .333\underline{/+105°}$$

$$I_{6-5} = I_{5-2} = 720\underline{/75°}$$

$$Z_{6-5} = \frac{V_{6-5}}{I_{6-5}} = \frac{240\underline{/+120°}}{720\underline{/+75°}} = .333\underline{/+45°}$$

Let's see that picture again:

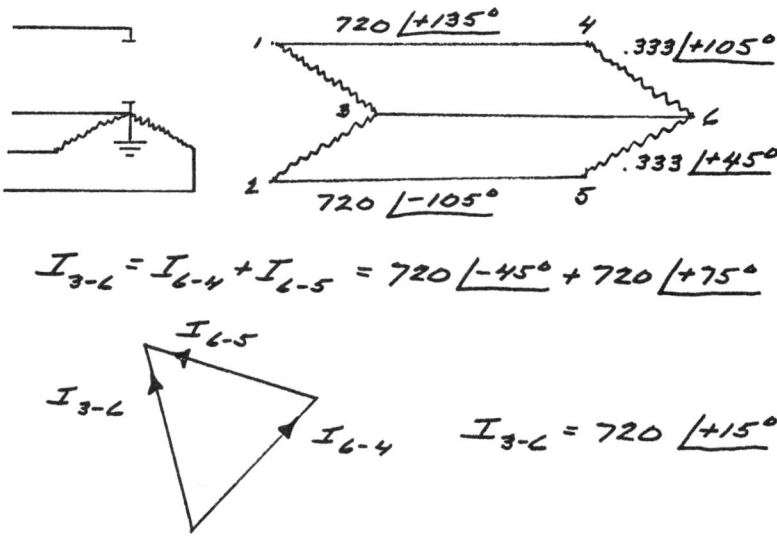

$$I_{3-6} = I_{6-4} + I_{6-5} = 720\underline{/-45°} + 720\underline{/+75°}$$

$$I_{3-6} = 720\underline{/+15°}$$

This checks with our previous calculations.

Now, let's cross check by looking at this thing another way. The equivalent impedance between points 4-6, and between points 6-5 is the parallel impedance of .577 ohms and .577 × 2, or 1.154 ohms.

$$\frac{.577 \times 1.154}{.577 + 1.154} = .385 \text{ ohms } \underline{/+75°}$$

But our equivalent impedances have been shifted 30°, Z_{6-5} forward, and Z_{4-6} back, to fit the requirements of our new two-phase circuit. So, equivalent impedances Z_{4-6} and Z_{6-5} are projections of the parallel impedance value on a vector 30° away, lying halfway between the two.

Checking, we see that:

.385 Cos 30° = .333 ohms.

Let's construct a vector diagram.

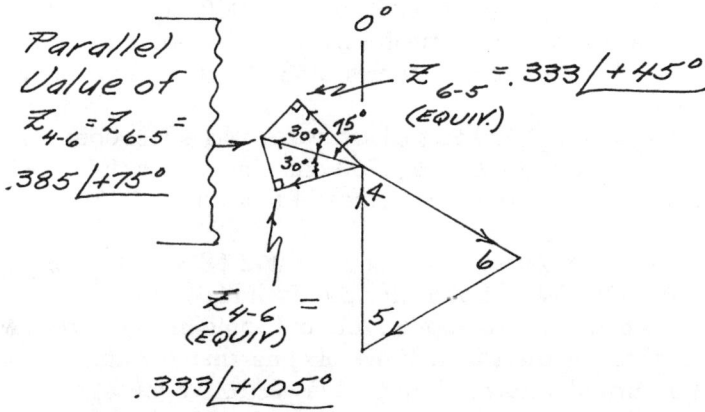

We have reduced the starting impedances of our three-phase motor to its two-phase equivalent.

Since an Open-Wye bank is only 87% efficient, we'd better use two 37½-kva transformers to supply this 50-horsepower motor.

Assume 2.5% impedance. If 2.5% of rated voltage is applied to the short-circuited unit, rated current will flow. Rated current on the high side is 37.5 ÷ 2.4 = 15.6 amperes. 2.5% of 2400 is 60 volts.

$$Z = V/I = 60/15.6 = 3.85 \text{ ohms.}$$

Now, let's break this down to R and X.

Full load loss is 400 watts. Core loss is 150 watts. Therefore, copper loss at rated current is 400—150 = 250 watts. This is $I^2 R$ at rated current.

$$R = \frac{250}{15.6^2} = 1.05 \text{ ohm.}$$

We then can determine X from our slide rule. Z, from the primary side = $3.85\underline{/74.1°}$ = 1.05 + j 3.70 ohms.

Now let's convert the two-phase equivalent motor impedance to its 2400-volt equivalent. Ten ohms at 240 volts allows 24 amperes to pass. This converts to 2.4 amperes at 2400 volts.

2400 V/2.4 amps = 1000 ohms. 1000 ohms/10 ohms = 100 = the square of the transformation ratio. So, our motor impedance of .333 ohms becomes .333 × 100 = 33.3 ohms at 2400 volts.

One of our 2400-volt phase wires was #1 copper, which has an impedance of .697 + j .750 ohms/mile. Our circuit is five miles long, so the total impedance of this wire is 3.485 + j 3.75 ohms.

The other phase is #6 copper at 2.18 + .829 ohms per mile, or 10.9 + j 4.145 ohms for the five-mile section.

The neutral is #6 copper, too, but some of the current will return through the earth. If we assume that one-third of the neutral current returns through the earth, we must adjust the neutral impedance to satisfy this condition. To do this, we must reduce the #6 copper neutral impedance by one-third, giving us an equivalent neutral impedance of 7.27 + j 2.75 ohms.

Now, let's put this information all down on a sketch of the system so we can visualize what we're doing.

We shall ignore substation impedance because it is smaller than the range of error which is probably already included by way of the assumptions we have had to make.

```
              LINE       TRANSFORMER    MOTOR
                        3.85/74.1°    33.3/105°
           3.485+j3.75  1.05+j3.7    -8.6+j32.1
  #1  ─────MMM──∩∩∩────MMM──∩∩∩──────MMM──∩∩∩──┐
                                                │
                                                │
                        3.85/74.1°    33.3/45°  │
           10.9+j4.145  1.05+j3.7    23.6+j23.6 │
  #6  ─────MMM──∩∩∩────MMM──∩∩∩──────MMM──∩∩∩──┤
                                                │
              7.78/20.7°                        │
              7.27+j2.75                        │
  #6 NT.─────MMM──∩∩∩───────────────────────────┘
```

Note that we could have connected our transformer so that the 105° impedance would have been tied to the #6 phase wire and the 45° impedance to the #1 phase wire. (We'll do that next.) Also, note that the motor presents a negative resistance to one phase.

When we look at this circuit we see that we shall have a current, I_1, flowing in the #1 copper phase, a current, I_2, flowing in the #6 copper phase, and a current, I_n flowing in the neutral. $I_1 + I_2 = I_n$, vectorially. $I_1 Z_1 + I_n Z_n = 2400\underline{/-120°}$. We know all of the Z's, but we can't find any of the I's. For instance, we can't just divide $2400\underline{/-120°}$ by $Z_1 + Z_n$, because I_1 doesn't flow through Z_n—I_n does.

We can't solve this one by the ordinary methods. It will require application of some logic and deduction.

First, let's add our impedances so we can compare them with each other:

#1 Phase:
    3.485 + j 3.750
    1.050 + j 3.700
   −8.600 + j 32.100
   −4.065 + j 39.550 = 39.8 /+95.9°

#6 Phase:
$$\begin{array}{r}10.900 + j\ 4.145\\ 1.050 + j\ 3.700\\ \underline{23.600 + j\ 23.600}\\ 35.550 + j\ 31.445 = 47.5\ \underline{/+41.5°}\end{array}$$

Neutral:
$$7.270 + j\ 2.750 = 7.78\ \underline{/+20.7°}$$

It is obvious that the bulk of the voltage drop will appear across the phase wires and motor coils combined and a relatively small voltage drop will appear along the neutral. It is also apparent that I_1 will exceed I_2 because Z_1 is smaller than Z_2. Also, $I_1 Z_1$ will be in the order of 2400 volts and will lie close to $\underline{/-120°}$. $I_n Z_n$ will be much smaller and may be pointed in almost any direction.

From this, we see that the scaler value of $I_1 Z_1$ will be close in value to $I_2 Z_2$. Let's assume, for the time being, that they are equal and see if we can find approximate values for I_1, I_2 and I_n.

$$I_1 Z_1 = I_2 Z_2$$

Then:

$$\frac{I_1}{I_2} = \frac{Z_2}{Z_1} = \frac{47.5}{39.8} = 1.2$$

Set I_2 equal to unity. I_1 lags V_1 by the impedance angle, neglecting the deflection caused by I_n, temporarily. Likewise, we'll assume for the present that I_2 lags V_2 by its impedance angle. Then:

$I_1 = 1.0\underline{/-215.9°}$ $(-120° - 95.9°)$

$I_2 = 1.2\underline{/+78.5°}$ $(+120° - 41.5°)$

Then $I_n = I_1 + I_2 = 1.0\underline{/+144.1°} + 1.2\underline{/+78.5°}$

$\qquad = -.810 + j\ .586 + .240 + j\ 1.18$

$\qquad = -.570 + j\ 1.766 = 1.86\underline{/+107.9°} = I_n$

$I_n Z_n = 1.86\underline{/+107.9°} \times 7.78\underline{/20.7°} = 14.5\underline{/+128.6°}$

Now, let's look at a vector diagram and see if we can get any helpful information:

UNIQUE POWER SYSTEM PROBLEMS—SOLVED

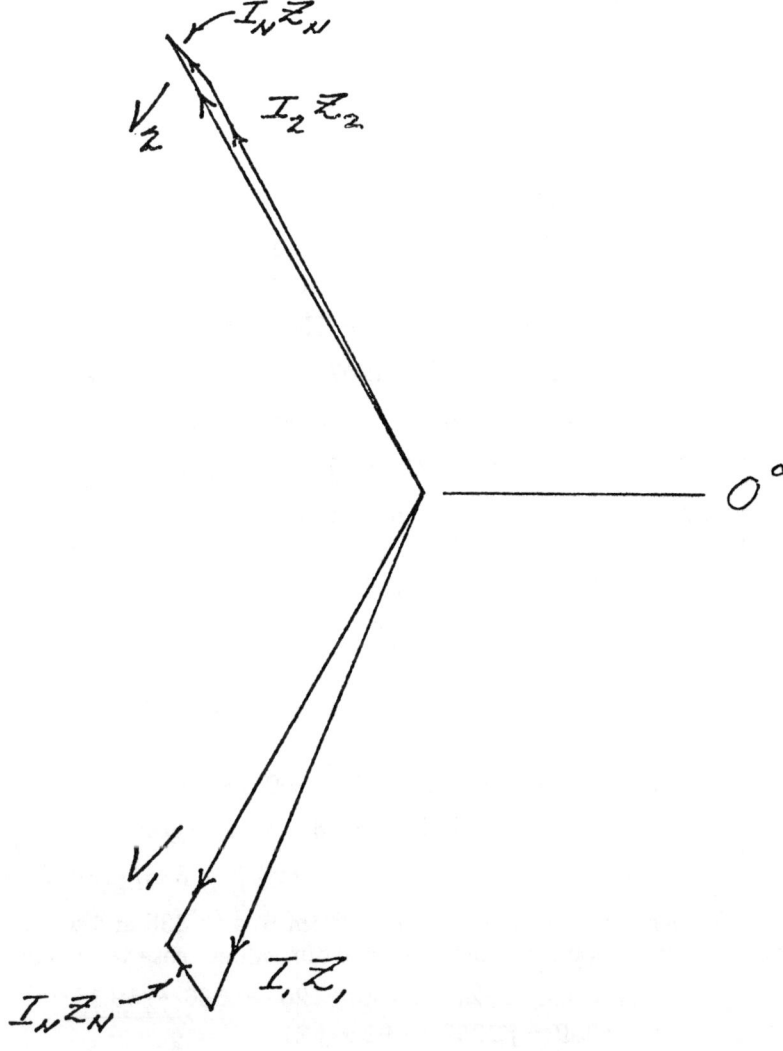

We know that $I_n Z_n$ is much smaller than $I_1 Z_1$ or $I_2 Z_2$, so we sketch it that way. $I_1 Z_1$ looks like it will be a little bigger than V_1 (2400 volts) and $I_2 Z_2$ will be smaller than V_2 by just about whatever $I_n Z_n$ turns out to be. Let's assume for a minute that $I_1 Z_1$ = 2400 volts at $\underline{/-120°}$. Then:

$I_1 Z_1 = 1.0/\underline{+144.1°} \times 39.8/\underline{95.9°} = 39.8/\underline{-120°}$

Then:
$I_1 Z_1 = 2400/C \quad C = \dfrac{2400}{39.8} = 60.3$

Then:

$I_n Z_n$ (full scale) $= 60.3 \times 14.5/\underline{+128.6°} = 874/\underline{128.6°}$

$V_1 - I_n Z_n = I_1 Z_1 = 2400/\underline{-120°} - 874/\underline{+128.6°}$

$= -1200 - j\,2080 + 545 - j\,683$

$= -655 - j\,2763 = 2840/\underline{-103.3°} = I_1 Z_1$

$V_2 - I_n Z_n = I_2 Z_2 = 2400/\underline{120°} - 874/\underline{+128.6°}$

$= -1200 + j\,2080 + 545 - j\,683$

$= -655 + j\,1397 = 1543/\underline{+115.1°} = I_2 Z_2$

Now we can determine the actual value of our trial currents:

$I_1 = \dfrac{I_1 Z_1}{Z_1} \quad \dfrac{2840/\underline{-103.3°}}{39.8/\underline{+95.9°}} = 71.4/\underline{+168.2°}$

$I_2 = \dfrac{I_2 Z_2}{Z_2} = \dfrac{1543/\underline{+115.1°}}{47.5/\underline{+41.5°}} = 32.5/\underline{+73.6°}$

$I_n = I_1 + I_2 = -69.9 + j\,14.6 + 9.2 + j\,31.2$

$\qquad = -51.5 + j\,45.8 = 68.9/\underline{+138.4°}$

$I_n Z_n = 68.9/\underline{+138.4°} \times 7.78/\underline{+20.7°} = 536/\underline{+159.1°}$

We see that $I_n Z_n$ has dropped from 874 to 536 and we are now ready to repeat the process until our values cease to change.

$V_1 - I_n Z_n = I_1 Z_1 = 2400/\underline{-120°} - 536/\underline{+159.1°}$

$= -1200 - j\,2080 + 492 - j\,212$

$= -708 - j\,2292 = 2399/\underline{-107.2°} = I_1 Z_1$

$V_2 - I_n Z_n = I_2 Z_2 = 2400/\underline{+120°} - 536/\underline{+159.1°}$

$= -1200 + j\,2080 + 492 - j\,212$

$= -708 + j\,1868 = 1998/\underline{+110.8°} = I_2 Z_2$

Note that $I_1 Z_1$ is less and $I_2 Z_2$ is more. We are closing in on the true figures. What current values derive from our new figures?

$$I_1 = \frac{I_1 Z_1}{Z_1} = \frac{2399/\!-\!107.2°}{39.8/\!+\!95.9°} = 60.3/\!+\!156.9°$$

$$I_2 = \frac{I_2 Z_2}{Z_2} = \frac{1998/110.8°}{47.5/\!+\!41.5°} = 42.0/\!+\!69.3°$$

I_1 is less, I_2 is greater than formerly.

$$I_n = I_1 + I_2 = -55.5 + j\,23.7 + 14.8 + j\,39.3$$
$$= -40.7 + j\,63.0 = 75.0/\!+\!122.9°$$

$I_n Z_n = 75.0/\!+\!122.9° \times 7.78/\!+\!20.7° = 584/\!+\!143.6°$

From 874 to 536 to 584. We're coming in for a landing!

$V_1 - I_n Z_n = I_1 Z_1 = 2400/\!-\!120° - 584/\!+\!143.6°$
$= -1200 - j\,2080 + 470 - j\,347$
$= -730 - j\,2427 = 2534/\!-\!106.7° = I_1 Z_1$

$V_2 - I_n Z_n = I_2 Z_2 = 2400/\!+\!120° - 584/\!+\!143.6°$
$= -1200 + j\,2080 + 470 - j\,347$
$= -730 + j\,1733 = 1880/\!+\!112.8° = I_2 Z_2$

$I_1 Z_1$ — from 2840 to 2399 to 2534. $I_2 Z_2$ — from 1543 to 1998 to 1880. We're obviously fluctuating around the true values and we're now getting close.

$$I_1 = \frac{I_1 Z_1}{Z_1} = \frac{2534/\!-\!106.7°}{39.8/95.9°} = 63.7/\!-\!202.6°$$

$$I_2 = \frac{I_2 Z_2}{Z_2} = \frac{1880/112.8°}{47.5/\!+\!41.5°} = 39.6/\!+\!71.3°$$

$I_n = I_1 + I_2 = -58.8 + j\,24.5 + 12.7 + j\,37.5$
$= -46.1 + j\,62.0 = 77.3/\!+\!126.6°$

$I_n Z_n = 77.6/\!+\!126.6° \times 7.78/\!+\!20.7° = 604/\!+\!147.3°$

158 UNIQUE POWER SYSTEM PROBLEMS—SOLVED

Only 20 change from the last value. Once more, and we'll be close enough for all practical purposes.

$$V_1 - I_n Z_n = I_1 Z_1 = 2400/\underline{-120°} - 604/\underline{+147.3°}$$
$$= -1200 - j\,2080 + 503 - j\,334$$
$$= -697 - 2414 = 2513/\underline{-106.1°} = I_1 Z_1$$

By golly, that one hardly changed at all!

$$V_2 - I_n Z_n = I_2 Z_2 = 2400/\underline{+120°} - 604/\underline{+147.3°}$$
$$= -1200 + j\,2080 + 503 - j\,334$$
$$= -697 + j\,1746 = 1880/\underline{+111.8°} = I_2 Z_2$$

$$I_1 = \frac{I_1 Z_1}{Z_1} = \frac{2513/\underline{-106.1°}}{39.8/\underline{+95.9°}} = 63.1/\underline{-202°}$$

$$I_2 = \frac{I_2 Z_2}{Z_2} = \frac{1880/\underline{+111.8°}}{47.5/\underline{+41.5°}} = 39.6/\underline{+70.3°}$$

$$I_n = I_1 + I_2 = -58.5 + j\,23.6 + 13.35 + j\,37.3$$
$$= -45.15 + j\,60.9 = 75.8/\underline{+126.55°}$$

$$I_n Z_n = 75.8/\underline{+126.55°} \times 7.78/\underline{+20.7°} = 590/\underline{+147.25°}$$

Our answers barely changed at all this time, so we know we've arrived.

Now, we can calculate the voltage drop on each phase of our equivalent two-phase motor impedance to see what voltage is available to start the motor.

$$I_1 Z_{m2} = 63.1/\underline{-202°} \times 33.3/\underline{+105°}$$
$$2101/\underline{-97°}$$
$$2101/2400 \times 100\% = 88\%\ (12\%\ \text{dip})$$
$$I_2 Z_{m2} = 39.6/\underline{70.3°} \times 33.3/\underline{+45°} =$$
$$1319/\underline{+115.3°}$$
$$1319/2400 \times 100\% = 55\%\ (45\%\ \text{dip})$$

Boy, look at that unbalance! And 45% dip on one phase. That is just too much dip and too much unbalance to accept.

What can we do to get better performance out of this set-up? Well, you probably saw it as soon as we listed the impedances in each phase. We will have better balanced phase impedances if we switch the 105° impedance from the #1 copper phase to the #6 copper phase and the 45° impedance from the #6 phase to the #1 phase. We do this by simply swapping the incoming 2.4 kv phases to the high side of our Open Wye-Open Delta bank. Our impedances are now:

#1 Phase:
$$\begin{array}{r} 3.485 + j\ 3.750 \\ 1.050 + j\ 3.700 \\ \underline{23.600 + j\,23.600} \\ 28.135 + j\,31.050 \end{array} = 41.9\ \underline{/47.8°} = Z_1$$

#6 Phase:
$$\begin{array}{r} 10.900 + j\ 4.145 \\ 1.050 + j\ 3.700 \\ \underline{-8.600 + j\,32.100} \\ 3.350 + j\,39.945 \end{array} = 40.0\ \underline{/85.2°} = Z_2$$

Neutral:
$$7.270 + j\ 2.750 = 7.78\ \underline{/20.7°}$$

Before, Z_1 was 39.8 and Z_2 was 47.5. Much more poorly balanced than our present figures of 41.9 and 40.0. We can expect much better balanced currents, voltage drops and dips.

We start again by assuming, for the time being, that the scaler value of $I_1 Z_1$ is equal to the scaler value of $I_2 Z_2$.

$$I_1 Z_1 = I_2 Z_2$$

$$\frac{I_1}{I_2} = \frac{Z_2}{Z_1} = \frac{40.0}{41.9} = .955$$

$I_1 = .955\underline{/-167.8°}$ $(-120.0° - 47.8°)$

$I_2 = 1.00\underline{/+34.8°}$ $(+120.0° - 85.2°)$

$I_n = I_1 + I_2 = .955\underline{/-167.8°} + 1.0\underline{/+34.8°}$

$\quad = .955\underline{/-132.0°}$

$\quad = -.933 - j\,.202 + .821 + j\,.571$

$\quad = -.112 + j\,.369 = .386\underline{/+107.0°}$

160 UNIQUE POWER SYSTEM PROBLEMS—SOLVED

$I_n Z_n = .386 / +107.0° \times 7.78 / +20.7° = 3.0 / +127.7°$

$I_1 Z_1 = .955 / -167.8° \times 41.9 / 47.8° = 40.0 / -120.0°$

$I_1 Z_1 \times C = 2400 \quad C = 2400/40 = 60.0$

$I_n Z_n = 3.0 / +127.7° \times 60.0 = 180.0 / +127.7°$

$V_1 - I_n Z_n = I_1 Z_1$

$2400 / -120° - 180.0 / +127.7° =$

$-1200 - j\,2080 + 110.1 - j\,142.4 =$

$-1090 - j\,2222.4 = 2475 / -116.1° = I_1 Z_1$

$V_2 - I_n Z_n = I_2 Z_2$

$2400 / +120° - 180.0 / +127.7° =$

$-1200 + j\,2080 + 110.1 - j\,142.4 =$

$-1090 + j\,1937.6 = 2223 / +119.4° = I_2 Z_2$

$I_1 = \dfrac{I_1 Z_1}{Z_1} = \dfrac{2475 / -116.1°}{41.9 / 47.8°} = 59.1 / -163.9°$

$I_2 = \dfrac{I_2 Z_2}{Z_2} = \dfrac{2223 / +119.4°}{40.0 / 85.2°} = 55.6 / +34.2°$

$I_n = I_1 + I_2 = -56.8 - j\,16.4 + 46.0 + j\,31.3$
$\qquad\qquad = -10.8 + j\,14.9 = 18.4 / +125.9°$

$I_n Z_n = 18.4 / +125.9° \times 7.78 / +20.7° = 143 / +146.6°$

$V_1 - I_n Z_n = I_1 Z_1$

$2400 / -120° - 143 / +146.6° =$

$-1200 - j\,2080 + 119 - j\,79$

$-1081 - j\,2159 = 2415 / -116.6° = I_1 Z_1$

$V_2 - I_n Z_n = I_2 Z_2$

$2400 / +120° - 143 / +146.6° =$

$-1200 + j\,2080 + 119 - j\,79 =$

$-1801 + j\,2001 = 2274 / +118.4° = I_2 Z_2$

UNIQUE POWER SYSTEM PROBLEMS—SOLVED

$$I_1 = \frac{I_1 Z_1}{Z_1} = \frac{2415/\underline{-116.6°}}{41.9/\underline{47.8°}} = 57.6/\underline{-164.4°}$$

$$I_2 = \frac{I_2 Z_2}{Z_2} = \frac{2274/\underline{+118.4°}}{40.0/\underline{85.2°}} = 56.9/\underline{+33.2°}$$

$I_n = I_1 + I_2 = -55.5 - j\,15.5 + 47.6 + j\,31.2$
$\quad = 7.9 + j\,15.7 = 17.6/\underline{+116.7°}$

$I_n Z_n = 17.6/\underline{+116.7°} \times 7.78/\underline{+20.7°} = 137/\underline{+137.4°}$

$V_1 - I_n Z_n = I_1 Z_1$

$2400/\underline{-120°} - 137/\underline{+137.4°} =$
$-1200 + j\,2080 + 102.3 - j\,94 =$
$-1097.7 - j\,2174 = 2435/\underline{-116.8°} = I_1 Z_1$

$V_2 - I_n Z_n = I_2 Z_2$

$2400/\underline{+120°} - 137/\underline{+137.4°} =$
$-1200 + j\,2080 + 102.3 - j\,94 =$
$-1097.7 + j\,1986 = 2269/\underline{+118.9°} = I_2 Z_2$

$$I_1 = \frac{I_1 Z_1}{Z_1} = \frac{2435/\underline{-116.8°}}{41.9/\underline{47.8°}} = 58.1/\underline{-164.6°}$$

$$I_2 = \frac{I_2 Z_2}{Z_2} = \frac{2269/\underline{+118.9°}}{40.0/\underline{+85.2°}} = 56.7/\underline{+33.7°}$$

There is no longer any appreciable change in our values, so they are now essentially correct.

We are now ready to send our phase currents through the equivalent motor coil impedances to determine the voltage which will be available to each.

$$I_1 Z_{m1} = 58.1/\underline{-164.56°} \times 33.3/\underline{+45°} =$$
$$1935/\underline{-119.6°}$$

$$1935/2400 \times 100\% = 80.7\%$$

$$I_2 Z_{m2} = 56.7\underline{/+33.7°} \times 33.3\underline{/+105°} =$$
$$1888\underline{/+138.7°}$$
$$1888/2400 \times 100\% = 78.7\%$$

We have 100—80.7 = 19.3% dip on the #1 copper phase and 100—78.7 = 21.3% dip on the #6 copper phase. Our motor will start just fine. The rural loads nearby will see a severe dip in voltage, but since the motor will start only occasionally, and only during flood season, it will not create an insupportable condition.

The value of this somewhat tedious problem does not lie so much in producing reasonably accurate values for the voltage dip we can expect when our 50-horsepower motor starts, as it does in demonstrating that just because a problem leads you into a dead-end where it appears there is no way out, is no reason to throw up your hands and quit. No, it is then time to put on your thinking cap and analyze the situation and develop a new line of attack. Don't be afraid to step out on your own and develop a new method to solve a problem. Just because your old textbooks don't have a similar problem with a standard solution is no sign *you* can't develop your own solution. Professors don't know everything!

INDEX

Accident, fatal, 4
Angle
   Impedance, 42, 145
   Little, 52
   Vertical, 22
Appliance damage, 23
Arrester, 78, 97
Autotransformer, 78

Bonded telephone cable, 2
"Black box" impedance, 83
Burndown, 97
Bus
   Capacity, 66
   Impedance of, 30
Capacitor, 2, 11, 37, 43, 53, 106
Circuit breaker, 3, 24, 96
Circulating current, 8
Clearance, mid-span, 55
Cluster mount, 136
Conductor
   Current carrying capacity, 35
   Impedance, 35
   Weight, 35
Conduit, 103
Current, circulating, 8

Data, 6, 30, 107, 113
Dead-end, 94
Delta-Delta, 14, 117
Delta-Wye, 115
Demand charge, 89
Diagram
   Equivalent transformer, 19
   Transformer connections
      Alternative, 136
      American standard, 118
Dip, voltage, 85, 144
Displaced neutral, 24, 51
Diversity factor, 43

Earth resistance, 94
Equivalent
   Delta, 67
   Impedance, 151
Eye, dilation of pupil, 28

Fault current, 66
   Blocked, 53
Flashover, 96
Floater, 58
Floating neutral, 6, 11, 53
Floodwall pump, 144
Flux, 104, 123
Freezing rain, 99
Fuse, blown, 2, 11, 53, 78, 95

Generator, 115
Ground
   Down, 94
   Driven, 95
   Effective, 3, 25
   Pole, 24
   Pole butt, 95

Heat of fusion, 99
Horizontal post construction, 97
Hotstick voltmeter, 55, 113
"H" structure, 58

Ice build-up, 99
Impedance
   Angle, 42, 145
   "Black box," 83
   Bus, 30
   Equivalent, 151
   Intermediate, 94
   Line, 30, 67
   Motor, 85, 145
   Parallel, 31, 83, 151
   Series Delta, 8

Starting, 151
System, 66
Transformer, 6, 25, 67
Infinite line, 94
Intermediate impedance, 94
Investment
   Annual cost, 13
   Postponed, 13
IR, 36
I^2R, 36, 99
IZ, 36

Leakage reactance, 18
Lighting load, 6
Lightning, 6, 25, 94
Lights, flare-up, 24
Load
   Factor, 89
   Lighting, 44
   Motor, 43
   Noncoincident peak, 91
   Point, 42
Loss
   Copper, 89
   Core, 89
   Factor, 91
   Line, 51
   Power, 32
   Transformer, 88
   Value, 1 kw on peak, 12

Motor
   Choke down, 78, 87
   Coil impedance, 149
   Cross-the-line start, 81, 144
   Impedance, 85, 145
   Overheating, 75
   Rotation, 114
   "Standing still" impedance, 86
   Starting, 80
   2300 volt, 78

Neutral
   Displaced, 24, 51

Excessive current, 3
Floating, 6, 11, 53
Grounded, 8
Shift, 25, 107
Voltage on, 3

Ohmmeter, 98
Ohm's Law, 3
One-point metering, 88
Open Delta, 143
Open Wye-open Delta, 4, 64, 143, 149
Overexcitation, 65
Overvoltage, 24
Ozone, 96

Paralleling subs, 116
Percentage, 36
Phase
   Rolling, 55
   Sequence, 143
Phasing
   Group I, 115
   Group II, 115
Phasing out, 19, 55, 113
Polarity, 122
Potential build-up, 96
Power factor, 42, 89
Profile, 22

Rate, 88
Reactance, leakage, 18
Recloser, 3
Regulator, 11, 37, 43, 75
Relays
   Thermal, 75
   Undervoltage, 87
Resistance
   Earth, 94
   Zero ground, 94
Rotation
   Motor, 114
   Phase vectors, 121

Sequence, phase, 143
Shielded line, 96
Sine wave, 118
Sleet, 99
"Slop," 92
Sparkover, 25, 96
Speed of electricity, 19
Stadia rod, 22
Starter, reduced voltage, 87
Starting impedance, 151
Surge, 98
Survey, 22
System impedance, 66

Taps, 8
Tariff, 88
Test system, 416 volt, 65
Thermal relay, 75
Transformer
    Connection diagrams, 117
    Equivalent diagram, 19
    Impedance, 6, 25, 67
    Loss, 88
Two phase and neutral system, 45, 144
Two-pole platform, 136

Unbalance
    Phase currents, 3, 107

Voltage, 75, 158
Undervoltage relay, 87
Unshielded line, 96

Vectors
    Arrow end, 118
    Rotation of, 121
    Voltage, 117
Voltage
    Decay, 50
    Dip, 85, 144
    Drop, 32, 36
    Impressed, 123
    Induced, 123
    Leftover, 11
    Overvoltage, 24
    Received, 50
    Reversed, 64
    Rise, 51
    Unbalance, 75, 158
    Vector, 118

Water table, 94
Wye-Delta, 6, 11, 15, 117
Wye-Wye, 117

Zero ground resistance, 94

"Z"-frame, 55